Springer Series on Bio- and Neurosystems

Volume 9

Series editor

The Springer Series on Bio- and Neurosystems publishes fundamental principles and state-of-the-art research at the intersection of biology, neuroscience, information processing and the engineering sciences. The series covers general informatics methods and techniques, together with their use to answer biological or medical questions. Of interest are both basics and new developments on traditional methods such as machine learning, artificial neural networks, statistical methods, nonlinear dynamics, information processing methods, and image and signal processing. New findings in biology and neuroscience obtained through informatics and engineering methods, topics in systems biology, medicine, neuroscience and ecology, as well as engineering applications such as robotic rehabilitation, health information technologies, and many more, are also examined. The main target group includes informaticians and engineers interested in biology, neuroscience and medicine, as well as biologists and neuroscientists using computational and engineering tools. Volumes published in the series include monographs, edited volumes, and selected conference proceedings. Books purposely devoted to supporting education at the graduate and post-graduate levels in bio- and neuroinformatics, computational biology and neuroscience, systems biology, systems neuroscience and other related areas are of particular interest.

More information about this series at http://www.springer.com/series/15821

Bhabesh Deka · Sumit Datta

Compressed Sensing Magnetic Resonance Image Reconstruction Algorithms

A Convex Optimization Approach

Springer

Bhabesh Deka
Department of Electronics
and Communication Engineering
Tezpur University
Tezpur, Assam, India

Sumit Datta
Department of Electronics
and Communication Engineering
Tezpur University
Tezpur, Assam, India

ISSN 2520-8535 ISSN 2520-8543 (electronic)
Springer Series on Bio- and Neurosystems
ISBN 978-981-13-3596-9 ISBN 978-981-13-3597-6 (eBook)
https://doi.org/10.1007/978-981-13-3597-6

Library of Congress Control Number: 2018963034

This Springer imprint is published by the registered company Springer Nature Singapore Pte Ltd.
The registered company address is: 152 Beach Road, #21-01/04 Gateway East, Singapore 189721, Singapore

Preface

This book presents a comprehensive review of convex optimization-based compressed sensing magnetic resonance image reconstruction algorithms. Compressed sensing MRI (CS-MRI) is successful in reducing the MRI scan time by two to five times. It takes only a few measurements in the frequency domain and then applies highly nonlinear recovery algorithms to reconstruct high-resolution MR images from the partial data. L1-norm-based optimization algorithms in convex optimization are popular for the reconstruction of MR images from partial Fourier data as they are stable, guarantee convergence at large scale, and are efficient, and standard solvers are readily available. Recently, fast convex optimization-based reconstruction algorithms have been developed which are quite competent to achieve the standards for the use of CS-MRI in clinical practice as well.

It is written primarily to have a quick review of the L1 optimization-based algorithms which are around for sparse representation in signal processing for quite a long time with their competitive performances and how they may be reconfigured to solve CS-MRI problem as well. At the outset, we have briefly described different approaches to solve CS-MRI problem in a convex optimization framework and then we move on to mention a few recently developed and fast algorithms in the field of CS-MRI. Toward the end, we try to analyze a few benchmarks for the evaluation of CS-MRI hardware/software, both qualitatively and quantitatively, and to standardize them for clinical practice. Finally, a glossary of CS-MRI applications in the field of bioinformatics and neuroinformatics is mentioned at both the research level and the clinical practice.

The contents in this book would allow postgraduate students, researchers, and medical practitioners working in the field of MR imaging to understand the need of CS-MRI and how it could revolutionize the soft tissue imaging technology to benefit the present healthcare scenario without making major changes in the existing scanner hardware. In particular, it is expected that researchers who have just stepped into this exciting field and would like to quickly go through the developments so far without diving into the detailed mathematical analysis would be benefited immensely. At the end, tracks of important future research directions are given which would help in consolidating this research for implementation in clinical practice.

Tezpur, India
June 2018

Bhabesh Deka
Sumit Datta

Acknowledgements

Dr. Bhabesh Deka would like to place on record his sincere thanks to Dr. K. R. Ramakrishnan, former Professor, Department of Electrical Engineering, IISc Bangalore, India, for introducing him to the topic 'Compressed Sensing' in signal processing during his short visit to the institute in Spring 2009 which motivated him deeply to pursue his Ph.D. in a closely related area later. He also wants to thank Dr. P. K. Bora, Professor, Department of Electronics and Electrical Engineering, Indian Institute of Technology Guwahati, India, for guiding him to pursue his Ph.D. in the topic 'Sparse Representations in Image Processing.'

Authors want to thank UGC, New Delhi, India, for giving a financial grant to the sponsored project on compressed sensing MRI at Tezpur University under the major research project scheme. They also want to thank Dr. S. K. Handique, Radiologist, GNRC Hospital Six-Mile Branch, Guwahati, India, for providing real 3D MRI datasets for simulations and helping them in interpreting relevant clinical information from the diagnostic images. Finally, thanks are equally due to the Department of Electronics and Communication Engineering, Tezpur University, for providing the necessary infrastructure to continue research in this exciting field.

Contents

1 Introduction to Compressed Sensing Magnetic Resonance Imaging 1
1.1 Introduction to MRI 1
1.2 MRI Data Acquisition 2
1.2.1 Single-Channel MRI 4
1.2.2 Multichannel (or Parallel) MRI 6
1.3 MR Image Contrast 6
1.3.1 Relaxation Time 6
1.3.2 Repetition Time 8
1.3.3 Echo Time 8
1.4 Types of MR Images 8
1.4.1 T_1-Weighted Image 9
1.4.2 T_2-Weighted Image 9
1.4.3 PD-Weighted Image 9
1.5 Compressed Sensing in MRI 11
1.6 Essentials of Sparse MRI 11
1.6.1 Sparsity of MR Images 12
1.6.2 Mutual Coherence 12
1.7 Design of CS-MRI Sampling Pattern 14
1.7.1 Variable Density Undersampling Pattern 16
1.7.2 Undersampling Pattern for Clinical MRI 16
1.8 Some Implementations of CS-MRI for Clinical Applications 18
1.9 Conclusions 21
References 21

2 CS-MRI Reconstruction Problem 23
2.1 Introduction 23
2.2 CS-MRI Problem Formulation 26
2.3 Conclusions 28
References 28

3 Fast Algorithms for Compressed Sensing MRI Reconstruction 31
3.1 Introduction 31
3.2 Operator Splitting Method 34
3.2.1 Iterative Shrinkage-Thresholding Algorithm 36
3.2.2 Two-Step Iterative Shrinkage-Thresholding Algorithm 38
3.2.3 Sparse Reconstruction by Separable Approximation 39
3.2.4 Fast Iterative Shrinkage-Thresholding Algorithm 40
3.2.5 Total Variation ℓ_1 Compressed MR Imaging 41
3.3 Variable Splitting Method 43
3.3.1 Augmented Lagrange Multiplier Method 44
3.3.2 Alternating Direction Method of Multipliers 47
3.3.3 Algorithm Based on Bregman Iteration 53
3.4 Composite Splitting 57
3.4.1 Composite Splitting Denoising 58
3.4.2 Composite Splitting Algorithm (CSA) 59
3.4.3 Fast Composite Splitting Algorithm (FCSA) 61
3.5 Non-splitting Method 61
3.5.1 Nonlinear Conjugate Gradient Method 62
3.5.2 Gradient Projection for Sparse Reconstruction 64
3.5.3 Truncated Newton Interior-Point Method 68
3.6 Conclusions 71
References 71

4 Performance Evaluation of CS-MRI Reconstruction Algorithms 75
4.1 Introduction 75
4.2 Simulation Setup 75
4.2.1 MRI Database Selection 76
4.2.2 Selection of Parameters 78
4.3 Performance Evaluation 79
4.4 Experiments on Convergence 86
4.5 Performance Evaluation of Iteratively Weighted Algorithms 88
4.6 Conclusions 97
References 97

5 CS-MRI Benchmarks and Current Trends 99
5.1 Introduction 99
5.2 Compressed Sensing for Clinical MRI 100
5.3 CS-MRI Reconstruction 100
5.3.1 k-Space Undersampling in Practice and Sparsifying Transform 101
5.3.2 Implementations 103
5.4 Image Quality Assessment 104
5.5 Computational Complexity 105

5.6 Current Trends 106
5.6.1 Interpolated CS-MRI (iCS-MRI) Reconstruction 106
5.6.2 Fast CS-MRI Hardware Implementation 106
5.7 Future Research Directions 108
5.8 Conclusions 109
References 109

6 Applications of CS-MRI in Bioinformatics and Neuroinformatics 111
6.1 Introduction 111
6.2 MRI in Bioinformatics 113
6.2.1 Whole-Body MRI 114
6.2.2 Magnetic Resonance Spectroscopy Imaging 115
6.2.3 Diffusion-Weighted MRI 115
6.2.4 Magnetic Resonance Angiography for Body Imaging 116
6.3 MRI in Neuroinformatics 116
6.3.1 Brain MRI 116
6.3.2 Functional MRI 117
6.3.3 Diffusion Weighted Brain MRI 118
6.3.4 Magnetic Resonance Angiography for Brain 118
6.4 Commercial CS-MRI Scanners 119
6.5 Conclusions 119
References 120

About the Authors

Dr. Bhabesh Deka has been Associate Professor in the Department of Electronics and Communication Engineering (ECE), Tezpur University, Assam, India, since January 2012. He is also Visvesvaraya Young Faculty Research Fellow (YFRF) of the Ministry of Electronics and Information Technology (MeitY), Government of India. His major research interests are image processing (particularly, inverse ill-posed problems), computer vision, compressive sensing MRI, and biomedical signal analysis. He is actively engaged in the development of the low-cost Internet of things (IoT)-enabled systems for mobile health care, high-throughput compressed sensing-based techniques for rapid magnetic resonance image reconstruction, and parallel computing architectures for real-time image processing and computer vision applications. He has published a number of articles in peer-reviewed national and international journals of high repute. He is also a regular reviewer for various leading journals, including *IEEE Transactions on Image Processing*, *IEEE Access*, *IEEE Signal Processing Letters*, *IET Image Processing*, *IET Computer Vision*, *Biomedical Signal Processing and Control*, *Digital Signal Processing*, and *International Journal of Electronics and Communications* (AEU). He is associated with a number of professional bodies and societies, like Fellow, IETE; Senior Member, IEEE (USA); Member, IEEE Engineering in Medicine and Biology (EMB) Society (USA); and Life Member, The Institution of Engineers (India).

Mr. Sumit Datta is currently pursuing his Ph.D. in the area of compressed sensing magnetic resonance image reconstruction in the Department of Electronics and Communication Engineering (ECE), Tezpur University, Assam, India. He received his B.Tech. in electronics and communication engineering from National Institute of Technology Agartala (NITA), Tripura, India, in 2011 and his M.Tech. in bio-electronics from Tezpur University in 2014. His research interests include image processing, biomedical signal and image processing, compressed sensing MRI, and parallel computing. He has published a number of articles in peer-reviewed national and international journals, such as *IEEE Signal Processing Letters, IET Image Processing, Journal of Optics, and the Multimedia Tools and Applications.*

Chapter 1
Introduction to Compressed Sensing Magnetic Resonance Imaging

Abstract Magnetic resonance imaging (MRI) is a widely used medical imaging tool where data acquisition is performed in the k-space, i.e., the Fourier transform domain. However, it has a fundamental limitation of being slow or having a long data acquisition time. Due to this, MRI is restricted in some clinical applications. Compressed sensing in MRI demonstrates that it is possible to reconstruct good quality MR images from a fewer k-space measurements. In this regard, convex optimization based ℓ_1-norm minimization techniques are able to reconstruct MR images from undersampled k-space measurements with some computational overheads compared to the conventional MRI where inverse Fourier transform is sufficient to get images from the fully acquired k-space. A few practical implementations of compressed sensing in clinical MRI demonstrate that they are able to significantly reduce the imaging time of traditional MRI. This is a very significant development in the field of medical imaging as it would improve both the patient care and the healthcare economy.

1.1 Introduction to MRI

Magnetic resonance imaging (MRI) is a rapidly growing medical imaging technique. It uses nonionizing radiation to generate high contrast images of soft tissues. This is also the key reason why MRI is highly recommended for soft tissue imaging. A large portion of our body is made up of water molecules which contains hydrogen atoms. The nuclei of hydrogen atoms have only one proton and they behave like the smallest entities of magnetism. When the body is placed in an external magnetic field then these little entities are aligned with and rotate around the axis of the magnetic field with an angular frequency proportional to the strength of the magnetic field. This frequency is known as the precessional or Larmor frequency [1, Ch. 14, p. 377]. Stronger the magnetic field, higher is the precessional frequency. Now, if a radio frequency (RF) pulse is also applied at the range of precessional frequency, then these spinning protons can absorb the RF energy and undergo transition from the lower energy state to the higher energy state, i.e., in parallel or antiparallel directions with respect to the applied magnetic field. At this stage, the spinning protons are said to be

B. Deka and S. Datta, *Compressed Sensing Magnetic Resonance Image Reconstruction Algorithms*, Springer Series on Bio- and Neurosystems 9,
https://doi.org/10.1007/978-981-13-3597-6_1

at resonance. When the RF pulse is turned off, the protons return back to the initial state and the difference of energy gives rise to the MR signal. The spinning protons from different tissues release energy at different rates because different tissues of the body have different chemical compositions and physical states.

The main limitation of MRI is due to the slow data acquisition process. An MR image consists of multiple acquisitions in k-space at intervals known as time of repetitions (TR). Each such acquisition is the result of the application of an RF excitation. However, these acquisitions are done in sequential form for a particular field of view (FOV) due to instrumental and physiological constraints. Therefore, it results in a long time for complete acquisition of the entire k-space to generate a single image. This slow imaging speed in MRI is quite a challenging task, especially for real-time MRI, like, dynamic cardiac imaging because then only a few samples can be collected during each cardiac cycle. In conventional MRI, k-space sampling follows the Nyquist criterion that depends on the FOV and the resolution of the MR image.

MR data acquisition can be accelerated by the use of high magnitude gradients since such gradients would minimize the TR time. However, the usage of such gradients with rapid switching is practically restricted as frequent variation of gradients would induce peripheral nerve stimulation in the patient. This fundamental speed limit of the MRI system has lead to the search for alternative viable technologies for enhancing the speed of MRI by undersampling the k-space without compromising the quality of reconstruction.

1.2 MRI Data Acquisition

Raw MRI data are stored in the form of a matrix in the k-space or the Fourier domain. Converting the k-space data into the image domain requires the 2D inverse Fourier Transform. In the k-space, the horizontal direction (k_x) is encoded by frequency encode gradient (FEG) and the vertical direction (k_y) is encoded by phase encode gradient (PEG). In either directions k_x or k_y, the frequency varies from $-f_{max}$ to $+f_{max}$ due to the respective gradient induced frequency variations. The center of the k-space represents the zero frequency. Traditionally, the k-space matrix is filled with one row at a time with the position of the row being determined by a particular pattern of the PEG. By slight variations of this gradient, different rows of the k-space matrix may be selected. The data in each row are obtained by applying the FEG in the k_x direction and thus by repeated applications of the same FEG, all the rows belonging to the entire k-space may be obtained.

We may summarize the MRI data acquisition steps as follows:

1. At first, a narrow RF pulse is applied along with the slice select gradient. The slice encode gradient (SEG) changes the precessional frequency of the target slice to the frequency of RF pulse so that the target slice can absorb energy from the RF

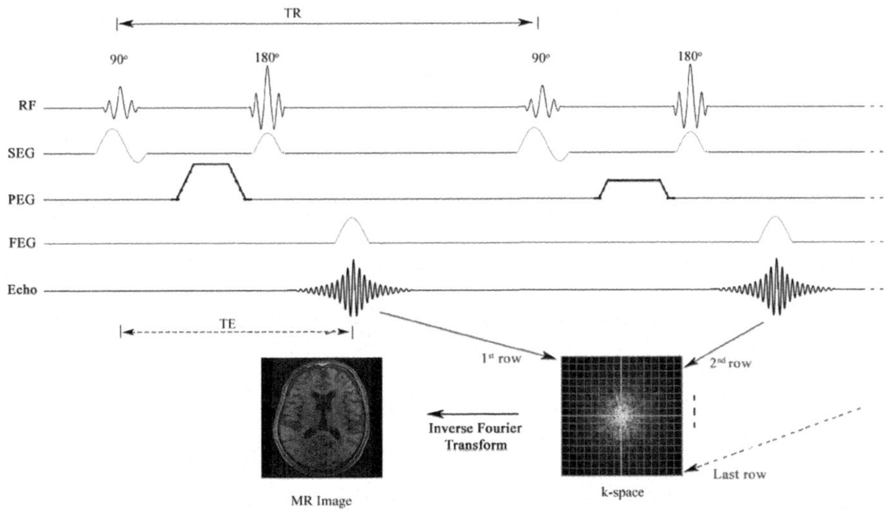

Fig. 1.1 Filling-up of a typical k-space matrix using repeated application of PEG and FEG pulse sequence

pulse. The amount of energy absorption depends on the magnitude and duration of the RF pulse.

2. Next, the PEG is applied for a brief duration to induce the phase difference information in the k-space data for localization of spatial information in the y-direction.
3. After a certain time called the echo time (TE), protons of the target slice start releasing energy which was absorbed during the RF excitation. During this period, the FEG is applied orthogonally to both the slice select and the phase encode gradients. This gradient induces the variation of frequency in the k-space data for localization of spatial information in the x-direction.
4. Then, a receiver coil along with an analog-to-digital converter (ADC) acquires the MR signal whose sampling rate depends on the bandwidth of the RF pulse. Acquired samples of the MR signal are stored row-wise in a 2D matrix representing the whole k-space.
5. Above steps are repeated several times with slight variations of the PEG to completely acquire the whole k-space. Finally, a 2D inverse Fourier transform converts the frequency domain information into the spatial domain which contains information of tissues of the selected slice.

Above process is pictorially summarized in Fig. 1.1. In Fig. 1.2, we pictorially demonstrate effects of PEG and FEG in the formation of the MR signal. The PEG produces spatial variations in angular frequencies of the excited protons whereas the FEG causes spatial variations in precessional frequencies of the spinning protons.

It is common to acquire the same k-space matrix repeatedly followed by simple averaging to increase the signal-to-noise ratio. Due to the repetition of RF pulses

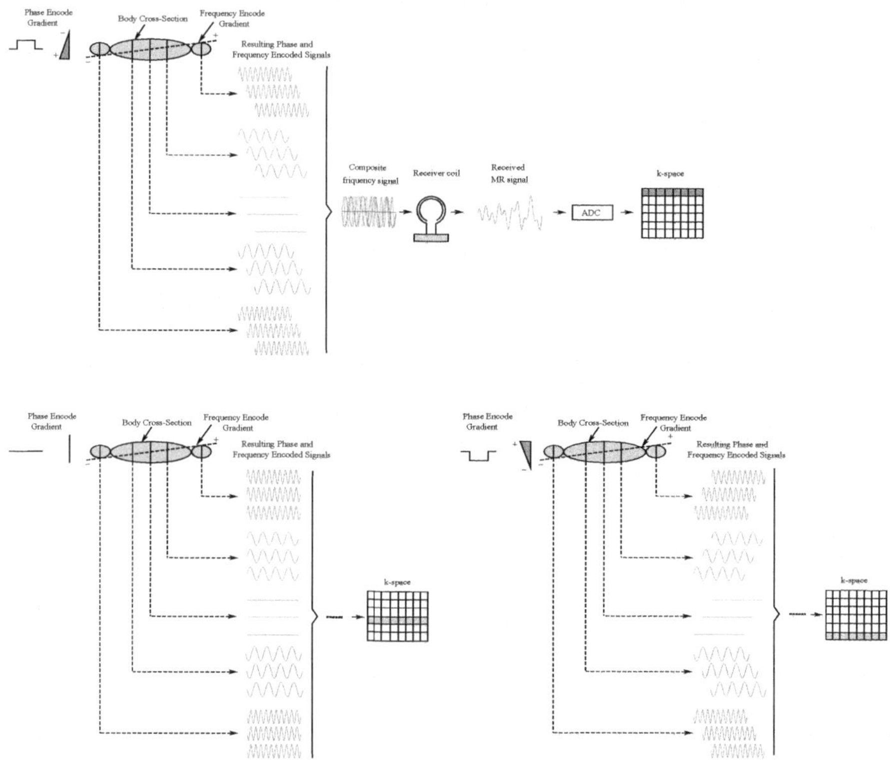

Fig. 1.2 Spatially dependent variation in the angular and the precessional frequencies of the protons due to the application of the PEG and the FEG

for several times to obtain different sets of k-space data, this acquisition process becomes a time-consuming one. Time taken by a 2D k-space data acquisition can be computed as—duration of a single TR $\times$ number of phase encode steps $\times$ number of signal averages. To reduce this data acquisition time, several attempts have been done and considerable modifications of the commercial MRI scanner are already been implemented. For example, by changing the pulse sequence, adding multiple receiver coils, etc. Depending on the number of receiver coils, the MRI scanners are divided into two categories which are discussed below.

1.2.1 Single-Channel MRI

Single-channel MRI is the simplest MRI modality in clinical diagnosis. Single receiver coil with uniform sensitivity over the entire FOV is used to acquire k-space data. It is further divided into two categories, namely, the single-echo and multi-echo

MRI. The first one is very common in routine diagnosis, whereas the second one is not so common. In MRI, contrast of various tissues in the image depends on the scanning parameters, i.e., TE and TR. For one set of parameters, a particular tissue, say, the gray matter may appear white while for another setting, the same may appear dark. Thus, there is one-to-many mapping between tissues and corresponding pixels of the image. Hence, MR images are not quantitative like computed tomography (CT) images. To overcome this drawback, multiple images are acquired with varying scanning parameters. After acquiring a set images, curve fitting could be done to find the best matching parameters to generate the desired tissue contrast. However, the process increases the scanning time of single-echo MRI significantly. To overcome this, the idea of multi-echo was developed where within the single TR, multiple echoes/images are collected for different sets of TE which help to obtain quantitative MR images within a reasonable time. Doctors also prefer multi-echo MR images to generate better contrasts between multiple tissues within the FOV because better tissue contrast makes the diagnosis easier [16, Chapter 2].

Several techniques are implemented to increase the data acquisition speed of single-channel MRI. Among them multi-slice data acquisition, techniques, like the fast spin echo (FSE), the echo planer image (EPI) acquisitions are most popular. Sequential slice by slice data acquisition concept is clinically unacceptable due to impractically long scan time. Multi-slice data acquisition significantly reduces the overall data acquisition time by exciting multiple slices within the same TR. The total number of slices excited within a single TR interval is TR $/(\text{TE} + C)$, where C is a constant and depends on the particular scanner and the overall scanning time is directly proportional to the number of excited slices.

In the FSE technique, application of multiple phase encode steps along with frequency encodes and 180° RF pulses are done within the same TR interval. These result in multiple k-space rows within the same TR interval. The amount of reduction in data acquisition time directly depends on the number of phase encode steps per TR interval. Using the FSE technique, one can achieve up to $16\times$ acceleration in clinical data acquisition. But, the main disadvantage of the FSE is that the SNR is reduced proportionately as the number of phase encode steps is increased within the same TR interval. It is also known as the *rapid acquisition with refocused echoes* (RARE) or *turbo spin echo*.

On the other hand, the EPI is extremely fast. There are mainly two types of EPI, namely, the single-shot and the multi-shot EPI. In the single-shot EPI, first one 90° RF pulse is applied followed by an initial PEG/FEG to start the data acquisition. Then, a 180° refocusing RF pulse is applied followed by continuous application of an oscillating FEG and PEG to acquire the whole k-space data in a zigzag manner corresponding to a selected slice. Here, images acquired with the single-shot EPI are generally having very poor resolution and low SNR. On the contrary, in the multi-shot EPI, instead of acquiring the entire k-space at one time it is done in segments where they are acquired with the application of multiple RF excitations unlike the single-shot EPI. This drastically improves the resolution and SNR [1, Chapter 15]

1.2.2 Multichannel (or Parallel) MRI

In single-channel MRI, reduction of scanning time by rapid switching of gradients also has physiological limitations as it may trigger peripheral nerve stimulations in the body. Using multiple receiver coils in parallel, the MR data acquisition time is reduced significantly. This approach is known as multichannel or parallel MRI (PMRI). Multiple receiver coils are placed at different positions of the scanner and they work independently to collect partial raw data of the target k-space. This process accelerates MRI scan time significantly compared to its single-channel counterpart. Data from each coil are then weighted differently and combined to give the full k-space data. The weighting factors of each receiver coil depend on the spatial sensitivity of the coil. The total scan time reduction directly depends on the number of independent receiver coils working in parallel [13].

For reconstruction of the full MR image from individual coil data, one needs information of the sensitivity profile of each receiver coil. In practice, it is never possible to obtain the exact sensitivity profile of a receiver coil which are usually determined through some time-consuming calibration processes. For this reason, image reconstruction from multiple receiver coils is still an active research area in PMRI.

Figure 1.3 shows an eight-channel head coil array PMRI. A channel or receiver coil is generally more sensitive to those signals emitted from tissues very close to it.

1.3 MR Image Contrast

Contrast of an image depends on the difference of its pixel intensities. In an MR image, each pixel corresponds to a voxel belonging to the corresponding FOV (or the target slice) whose values do not directly depend on tissues but on the scanning parameters, i.e., TR and TE. Next, we discuss different parameters influencing the MR image contrast.

1.3.1 Relaxation Time

The term relaxation means that the spinning protons are back to their equilibrium state. Once the radio frequency (RF) pulse is turned off, the protons will have to realign with the axis of the static magnetic field B_0 and give up all their excess energies to the surrounding environment [1, Chapter 14], [17]. The relaxation consists of two important features which could be described in terms of the following events in time:

T1 or Longitudinal Relaxation Time

T1 or the longitudinal relaxation time is the time taken for the spinning protons to realign along the longitudinal axis. It is also called the spin–lattice relaxation

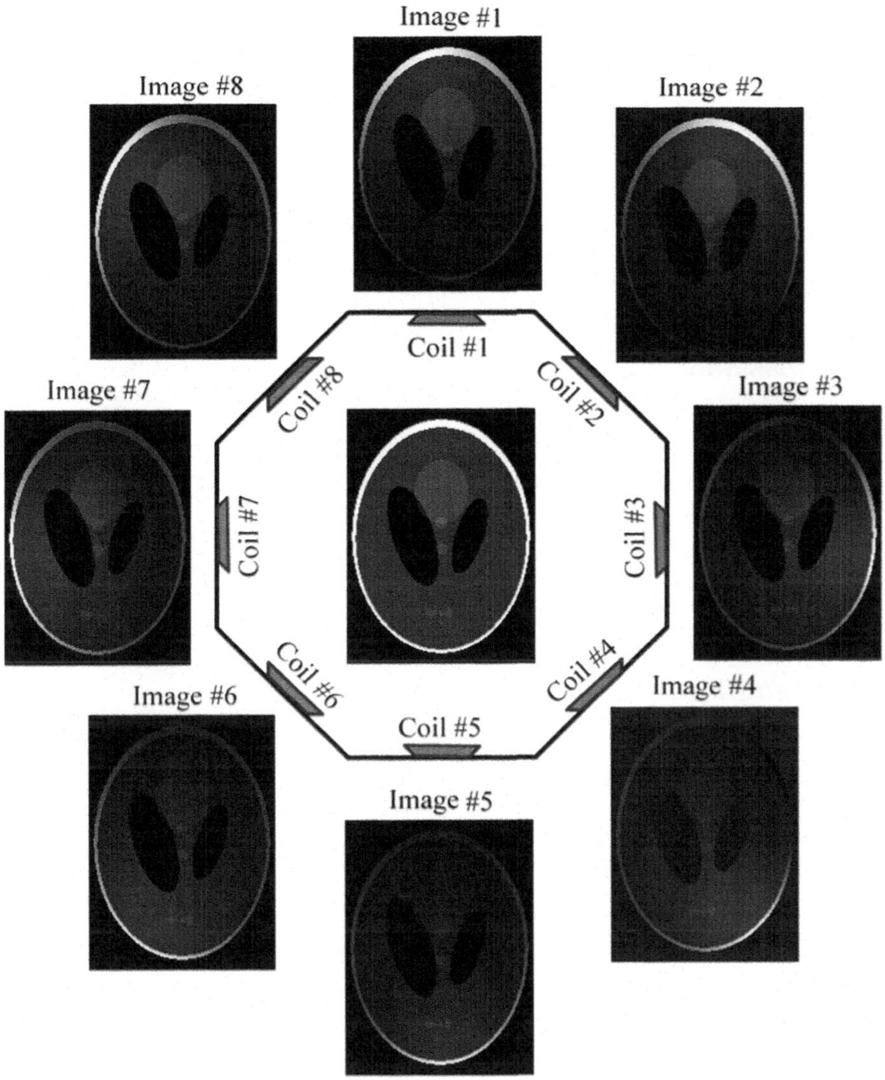

Fig. 1.3 An example of eight-channel head coil array

time because during this period, each spinning proton releases its energy which was obtained from the RF pulse back to the surrounding tissue (lattice) in order to reach their equilibrium states thus reestablishing the longitudinal magnetization again. It objectively refers to the time interval to reach up to 63% recovery of the longitudinal magnetization.

T2 or Transverse Relaxation Time

On the other hand, T2 or transverse relaxation time is the dephasing (loss of coherence) period of spinning protons due to the intrinsic magnetic properties of our body after the 90° RF pulse is turned off. It is also known as the spin-spin interactions time, during this period, the loss of transverse magnetization occurs. Specifically, T2 or transverse relaxation time refers to the time interval from the time of occurrence of the peak level of transverse magnetization to the time when it decays up to 37% from its peak value.

1.3.2 Repetition Time

The repetition time (TR) is the interval between the application of two adjacent (90°) RF excitation pulses. It determines recovery of longitudinal magnetization after each excitation pulse. For example, if we set short TR, the tissue having long T1 time like the CSF will appear dark and the tissue having short T1 time like the fat will appear bright.

1.3.3 Echo Time

Application of the (90%) RF excitation pulse produces maximum transverse magnetization and when it gets turned off, spinning protons start dephasing. After a delay of half of the *echo time* (TE), i.e., the TE/2, an (180°) RF pulse is applied to inverse the spin and rephase the transverse magnetization. This rephasing produces a measurable signal in the receiver coil after another TE/2 following the application of the (180°) RF pulse. Specifically, the *echo time* (TE) refers to the time interval between the application of (90°) RF excitation pulse and the occurrence of the peak of the echo signal in the receiver coil. If we set long TE, the tissue having long T2 time like the CSF will appear bright and the tissue having short T2 time like the fat will appear dark.

1.4 Types of MR Images

Depending on the proton density (PD), i.e., number of hydrogen atoms in a unit volume, T1, and T2 relaxation times, MR images are classified as PD, T1, and T2 weighted images, respectively. Selection of TR and TE parameters for getting T1, T2, and PD-weighted images is summarized in Table 1.1. In the following, we briefly mention their formation in any typical MRI scanner.

1.4.1 T_1-Weighted Image

An image contrast which mainly depends on the T1 or longitudinal relaxation time is called the T1-weighted image. Here, both TR and TE times are kept short, in the ranges of 400–600 ms and 5–30 ms, respectively. Tissues with a short T1 appear bright because they recover most of their longitudinal magnetization during the TR interval and thus produce a stronger MR signal. Tissues with a long T1 appear dark because they do not recover much of their longitudinal magnetization during the TR interval and thus produce a weaker MR signal.

1.4.2 T_2-Weighted Image

In a T2-weighted image, the contrast mainly depends on the T2 or transverse relaxation time. Here, both the TR and the TE are kept long, in the ranges of say, 1,500–3,500 ms and 60–150 ms, respectively. Tissues with a short T2 appear dark on T2-weighted images because they lose the transverse magnetization quickly. On the other hand, tissues with a long T2 appear bright on T2-weighted images because for them, transverse magnetization would decay very slowly.

1.4.3 PD-Weighted Image

An image with contrast mainly depending on the number of protons or hydrogen atoms per unit volume of tissue is called the PD-weighted image. A long TR and very short TE are selected to reduce effects of T1 and T2 on the image. Commonly, the TR and the TE are kept in the ranges of 1,500–3,500 ms and 5–30 ms, respectively. Here, differences of signals magnitudes received from different tissues depend only on the density of protons of the corresponding tissue volume. Signal intensities of a few selected tissues on T1, T2, and PD-weighted MR images are shown in Table 1.2. Some other types of MR images commonly used in clinical diagnosis are: (a) gradient-echo T1-weighted images like, FLASH (Siemens), SPGR (General Electric), T1-FFE (Philips), RF-FAST (Marconi); (b) gradient-echo T2*-weighted images; (c) gradient-echo PD-weighted images; (d) Short T1 Inversion Recovery (STIR) images; and (e) FLuid Attenuated Inversion Recovery (FLAIR) images [17].

MRI has a clear edge over other medical imaging techniques for clinical imaging as mentioned earlier due to high contrast for soft tissue imaging, noninvasiveness, no ionizing radiation, etc. However, some instrumental limitations, like, high gradient amplitude, and slew rate; physiological problems, like, peripheral nerve stimulation (PNS) limits its data acquisition speed. As a result, data acquisition becomes comparatively slow. As it is very difficult to stay a long time, depending on the FOV, in a claustrophobic environment inside the bore of a scanner without any body movement.

Table 1.1 Image contrast for different repetition time and echo time

MR image	Repetition time (TR)	Echo time (TE)
T1-weighted	Short	Short
T2-weighted	Long	Long
PD-weighted	Long	Short

Table 1.2 Signal intensities of different tissues on T1 and T2-weighted images

Tissue	T1-weighted image	T2-weighted image	PD-weighted image
Bone	Dark	Dark	Dark
CSF	Dark	Bright	Bright (darker than T2)
Cyst	Dark	Bright	Dark
Fat	Bright	Bright	Bright (slightly darker)
Gray matter	Gray	Gray	Bright gray
Muscle	Gray	Gray	Gray
White matter	Brightish	Darkish	Gray
Tumor	Dark	Bright	Dark

This is particularly a serious concern for persons having diseases or in emergency conditions. So, reduction of MRI scan time is of paramount importance for making MRI highly preferable for diagnostic imaging.

Clinical MRI scanners having magnetic field intensities of 1.5–3 T are very common. In MRI, higher magnetic fields increase the SNR, for example, a 3 T scanner offers two times improvement in SNR over a 1.5 T scanner. Also, 4× acceleration can be achieved for the same SNR value, like, 28 s breath-hold abdominal imaging can be done in 7 s only, which is quite comfortable for a patient [24]. But, it has also practical limitations, like, maintaining the magnetic field homogeneity which is the major production challenge. It increases size and weights of the scanner, for example, a 4.0 T scanner required 20 tones of cryogens for cooling and 150 tones of steel shielding. Moreover, higher magnetic fields require stronger and faster gradient switching which causes acoustic noise apart from cardiac and peripheral nerve stimulations.

To reduce the data acquisition time, several attempts have been made in the past and recent times which may be broadly categorized into two, namely, hardware- and software-based approaches. Multichannel parallel MRI is the most well-known hardware-based approach. They are economically expensive and require considerable time in development. On the other hand, software-based approaches do not require any modification of the existing scanner hardware. A controlling computer is just sufficient to change the imaging pulse sequences and then reconstruct images using some postprocessing techniques which may be done offline. In this direction, Lustig et al. [14, 15] introduces a new paradigm of sampling in MRI where the signal is also compressed while it is acquired, which is popularly known by "compressed sensing" among the signal processing community. He coined the term compressed sensing MRI (CS-MRI). It is quite possible to reconstruct very good diagnostic quality MR images from just 20–30% k-space data using the theory of CS-MRI. This is a major breakthrough for the development of rapid MRI in clinical applications.

1.5 Compressed Sensing in MRI

The *compressed sensing or compressive sensing, or compressive sampling* (CS) is the process of efficiently acquiring either a sparse signal directly or a signal which is compressible in some known domain. Details of the CS principles are reported in [3]. The two key requirements for the success of the CS theory are: (1) signal or image should be either sparse or compressible in some transform domain, and (2) the aliasing artifacts due to the random under sampling must be incoherent in the sparse representation domain.

In MRI, undersampling in k-space by periodic interleaving leads to the violation of the Nyquist criterion for signal/image sampling. This produces aliasing artifacts in the reconstructed signal as observed in Fig. 1.6c. In order to mitigate this problem, first, the undersampling artifacts are desired to be incoherent or less visible in the reconstructed images and second, the signal must have either a spatial or temporal redundancy. MRI, in particular, has spatial redundancy. It means that in the transform domain (like wavelets), only a few elements contain maximum information of the image. Only these coefficients are enough for reconstruction of the original image without any significant loss of information. According to Candes et al. [3], a signal may be reconstructed from a fewer samples or measurements than the Shannon–Nyquist theorem if the signal is sparse or at least compressible in some transform domain.

MRI data acquisition process is performed in k-space or Fourier plane. MR images are sparse in the wavelet domain and a strong incoherence between the Fourier and the wavelet domains exists. Thus, we see that MRI inherently fulfills the key requirements of the CS [15].

1.6 Essentials of Sparse MRI

One can accurately reconstruct an MR image by acquiring only a few random samples of the k-space rather than the whole k-space provided it satisfies the key requirements of the CS and any nonlinear reconstruction scheme is able to enforce sparsity of the MR image in the transform domain together with the consistency of data acquired in the k-space [14, 16].

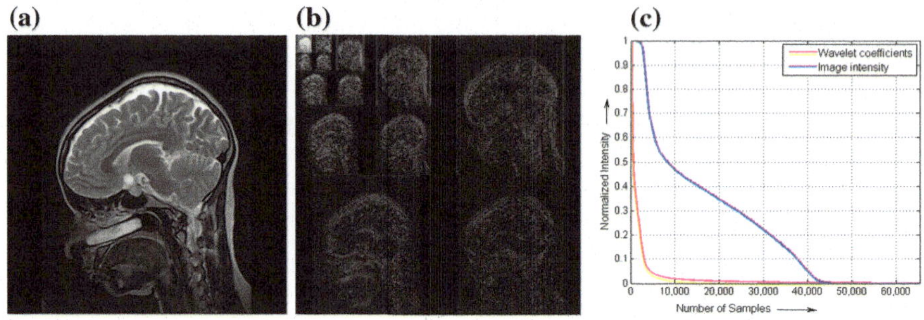

Fig. 1.4 Sparse representation of MR image in transform domain. **a** Brain MR image, **b** sparse representation of MR image in wavelet domain and **c** comparison of normalized intensity of the wavelet coefficients and the image pixels

1.6.1 Sparsity of MR Images

Generally, the natural signals or images are not sparse in its own domain. But, they are compressible in some transform domain. Suppose, $\mathbf{x}$ is an $\sqrt{n} \times \sqrt{n}$ image which is dense in spatial domain but compressible when projected to an appropriate basis set Ψ (like the discrete cosine transform, the wavelet, etc.), as shown in Fig. 1.4. If we now reorder $\mathbf{x}$ lexicographically to form a column vector of dimension $n \times 1$, then we can express it as $\mathbf{x} = \Psi\boldsymbol{\alpha}$, where $\Psi \in \mathbb{R}^{n\times n}$ is called the transform matrix or dictionary and $\boldsymbol{\alpha} \in \mathbb{R}^n$ are known as the sparse transform coefficients.

1.6.2 Mutual Coherence

For the accurate reconstruction of a signal or image from the CS data, the sensing matrix Φ must be incoherent with the sparse representation/transform basis set Ψ [4]. In the CS-MRI literature, the sensing matrix is represented by $\mathbf{F}_u$ such that $\mathbf{F}_u = \Phi\mathbf{F}$, where $\Phi \in \mathbb{R}^{m\times n}$ is a binary matrix with each row having all zeros except a single "1" for randomized selection of a row from the $(n \times n)$ discrete Fourier transform matrix $\mathbf{F}$. Now, suppose an MR image $\mathbf{x}$ is acquired with a sensing matrix $\mathbf{F}_u \in \mathbb{R}^{m\times n}$ such that the acquired signal $\mathbf{y} = \mathbf{F}_u\mathbf{x} \in \mathbb{R}^m$ and $m \ll n$. The mutual coherence between the sensing basis $\mathbf{F}_u$ and the representation basis Ψ can be defined mathematically as

$$\mu(\mathbf{F}_u, \Psi) = \sqrt{n} \max_{1\le k, j\le n} \left|\left\langle (\mathbf{F}_u)_k, \Psi_j \right\rangle\right| \tag{1.1}$$

whose value lies in the range $\left[1, \sqrt{n}\right]$ [3]. If $\mathbf{F}_u$ and Ψ contain correlated elements then incoherence is small. For reconstruction with less aliasing artifacts, incoherency between $\mathbf{F}_u$ and Ψ should be large as much as possible.

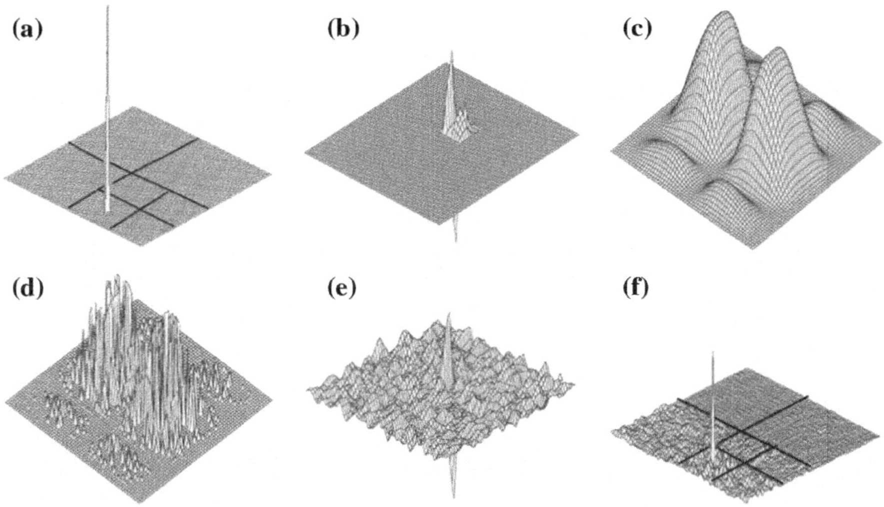

Fig. 1.5 TPSF in the wavelet domain due to the variable undersampling Fig. 1.8b in Fourier domain. **a** A single point in the ith location of wavelet domain, **b** the corresponding image domain and **c** the transformed Fourier (k-space) representation of (**b**). **d** The undersample k-space with sampling scheme Fig. 1.8b, **e** and **f** are the corresponding image and wavelet domain representation of **d**, respectively

The point spread function (PSF) is another tool to compute incoherence [14]. In transform domain, the incoherence is measured by transform point spread function (TPSF). TPSF measures the influence of a point in the ith location to another point in the jth location of the transform domain. It is expressed mathematically by

$$\text{TPSF}(i;\, j) = e_j^* \Psi F_u^* F_u \Psi^* e_i, \tag{1.2}$$

where $\mathbf{e}_i$ is the vector to represent the unit intensity pixel at ith location and zero elsewhere. If we sample the k-space according to the Nyquist rate then there will be no interference in transform domain, i.e., $\text{TPSF}(i;\, j)_{i \neq j} = 0$. But, undersampling in k-space domain causes the interference in transform domain, i.e., $\text{TPSF}(i;\, j)_{i \neq j} \neq 0$. Figure 1.5 shows the incoherent interference in the wavelet domain due to the random undersampling in k-space.

Figure 1.5a shows a unit intensity coefficient in the wavelet domain at the ith position. Suppose, this coefficient is transformed back to the image domain by taking the inverse wavelet transform. Then, we can take its Fourier transform to show the incoherency between the wavelet and the Fourier transforms. This is observed in Fig. 1.5c. Now, if the representation in Fourier domain is randomly undersampled and transformed back, first to the image domain and then to the wavelet domain, it is observed that the energy of the coefficient in the wavelet domain spreads mainly near the same scale and orientation and these are incoherent with unit intensity coefficient.

For successful recovery by the CS, we must see that the measurement matrix $\mathbf{F}_u\Psi$ closely follows a very important property called the *restricted isometry property* (RIP) which is explained next.

Restricted Isometry Property

Candès and Tao [2] define the so-called restricted isometry property (RIP) as follows: If $\mathbf{A}$ is the measurement matrix, then it would satisfy the RIP of order s with $s \ll n$, if there exists an isometry constant $0 < \delta_s < 1$, such that for all s-sparse vector, $\mathbf{x}$

$$(1 - \delta_s)\,||\mathbf{x}||_2^2 \leq ||\mathbf{A}_\Lambda \mathbf{x}||_2^2 \leq (1 + \delta_s)\,||\mathbf{x}||_2^2, \tag{1.3}$$

where $\mathbf{A}_\Lambda$ is the $m \times |\Lambda|$ submatrix of $\mathbf{A}$ with $|\Lambda| \leqslant s$ and δ_s is the smallest number that satisfies Eq. 1.3. Orthogonal matrix has $\delta_s = 0$ for all s. $\delta_s < 1$ allows for reconstruction of any signal $\mathbf{x}$. If $\delta_s << 1$, the matrix $\mathbf{A}$ has a large probability to accurately reconstruct the signal $\mathbf{x}$. Computational complexity to estimate δ_s is very high because the estimation problem is combinatorial in nature. However, if $\mathbf{A}$ is constructed using Gaussian or Bernoulli random variables and $m \geq C.s.\log(n/s)$ for some constant C, the RIP is fulfilled with extremely high probability. Similarly, for a randomly undersampled Fourier matrix, the RIP constant δ_s is more restricted and observed that the corresponding number of measurements $m \geq C.s.\log^4(n)$ [5].

1.7 Design of CS-MRI Sampling Pattern

In 2D MRI, two gradients, namely, the phase encode gradient and the frequency encode gradient are used to acquire the particular FOV. The sampling along the frequency encode direction is not a limiting factor in terms of scan time, i.e., in 2D MRI one requires to implement only one-dimensional undersampling. Therefore, in 2D MRI, the k-space undersampling can be done in the phase encode direction only by using parallel line trajectories as shown in Fig. 1.9a. In 3D MRI one extra gradient, namely, the slice select gradient is required to acquire the particular FOV. In 3D MRI, two-dimensional undersampling is applicable, i.e., undersampling is done in both phase encode and slice select directions but frequency encode direction is fully acquired as in 2D MRI. This is observed in Fig. 1.9e. In case of dynamic MRI, time is added as an additional dimension with two or three k-space gradients, where undersampling is performed in frequency versus time domain. In our discussions, we are assuming only Cartesian trajectories, since most of the clinical MRI use them.

To use the CS in practical MRI, one needs to design an efficient undersampling pattern (encoder) which ensures that measured data contain almost all information of the original image and a good reconstruction algorithm (decoder), which can recover the encoded information from the measured data [26].

Data acquisition is the most important part in compressed sensing MRI. The main target is how efficiently one can acquire only a few samples for reconstruction of the image without compromising the quality of the image. Incoherent aliasing artifact

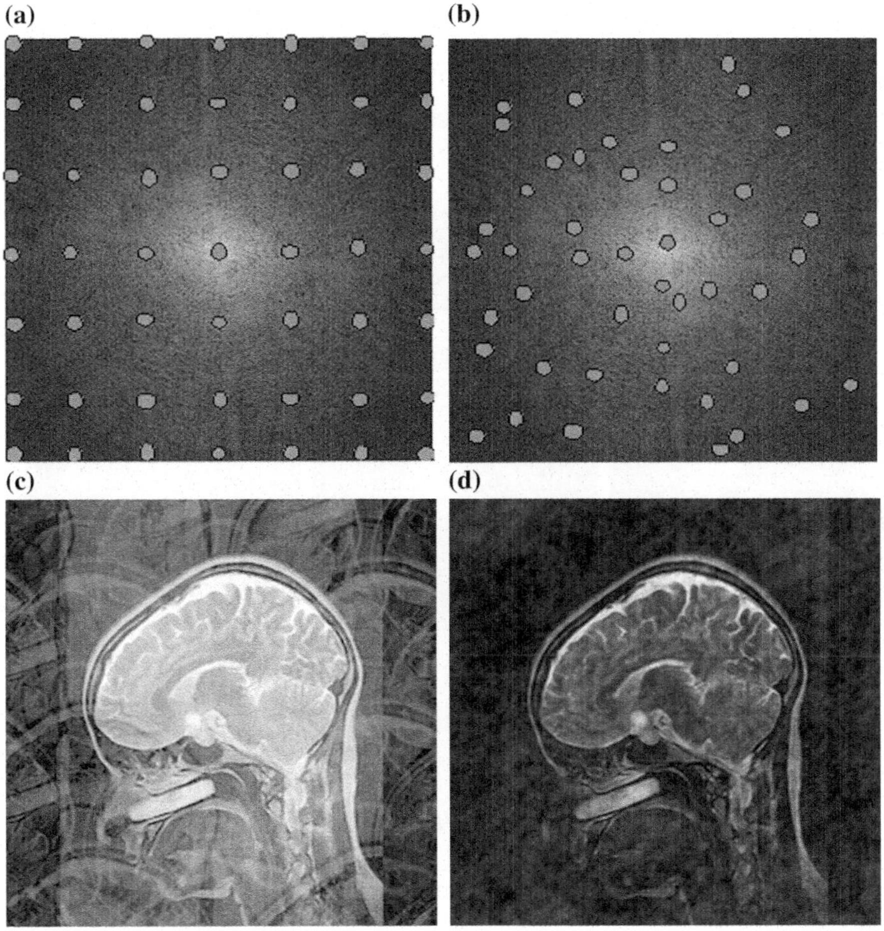

Fig. 1.6 Different k-space sampling and the corresponding MR images. **a** the uniform k-space undersampling, **b** the random k-space undersampling, **c** and **d** are the zero-padded inverse Fourier transform of (**a**) and (**b**), respectively

is an important criterion for CS reconstruction. Equispaced k-space under sampling creates coherent aliasing artifacts as shown in Fig.1.6. In case of coherent aliasing artifacts reconstruction of the original image is not possible, as the actual reconstructed image would overlap in an indistinguishable manner with its aliases. But, in the case of random undersampling, situation is different. Although the reconstructed image contains artifacts but they are incoherent and very much similar to white noise as shown in Fig.1.6. These artifacts appear in the MR image due to the leakage of energy caused by random undersampling. This type of random noise like artifacts can be easily removed during the reconstruction process by thresholding or any other process [14, 15].

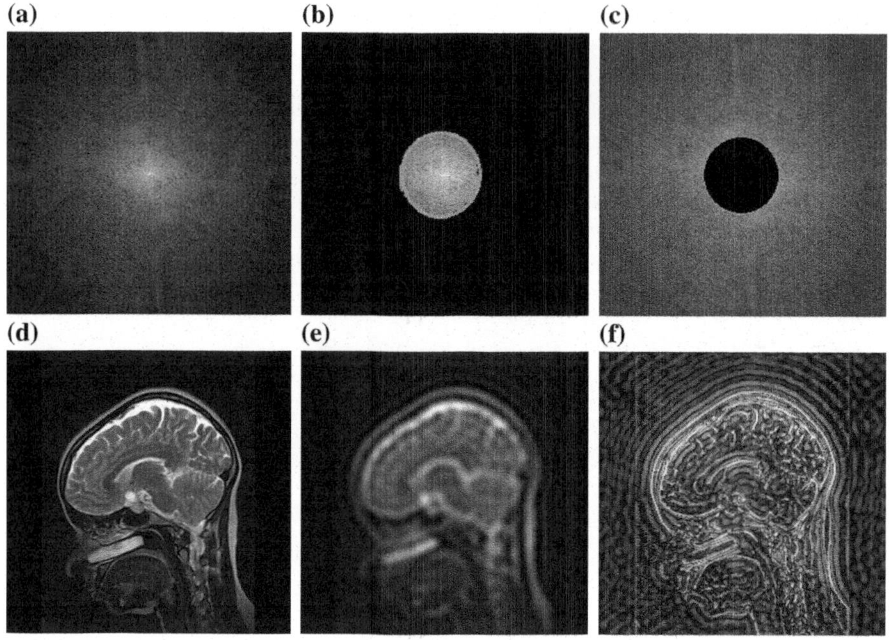

Fig. 1.7 Different k-space data and corresponding MR images. **a** The whole k-space of a brain MR image, **b** only the center region of k-space, **c** only the periphery region of the k-space, **d**, **e**, and **f** are the corresponding MR images of (**a**), (**b**) and (**c**), respectively

1.7.1 Variable Density Undersampling Pattern

An MR image is acquired in k-space. k-space has some unique properties. Center region of the k-space contains information about gross structure and contrast of the underlying image that is essential to reproduce it. On the other hand, the peripheral region contains the high-frequency information needed for better spatial resolution. Therefore, if we have to undersample the k-space, we have to acquire more samples from the center region and relatively less samples from the periphery. Figure 1.7 clearly shows that the center region of the k-space contains more information about the underlying MR image. Therefore, variable density undersampling scheme would be a better choice for acquiring samples from the k-space [14, 22]. Researchers proposed various variable density sampling patterns based on different techniques [9, 14, 23]. Figure 1.8 shows different variable density sampling patterns.

1.7.2 Undersampling Pattern for Clinical MRI

Parallel imaging (PI) sufficiently improves the scan time of clinical MRI. However, in some cases like pediatric subjects and 3D MR imaging, we need faster data acqui-

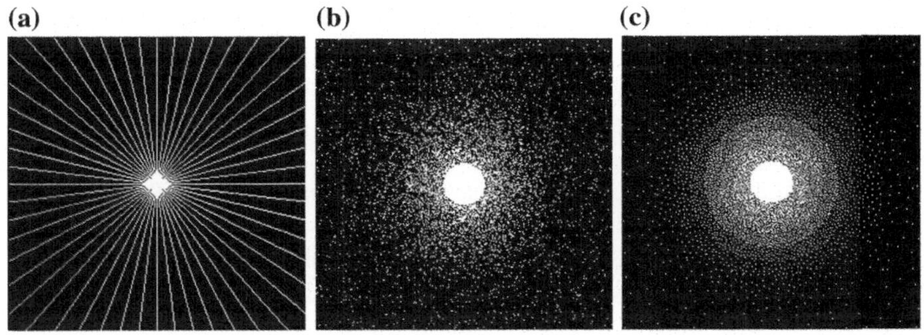

Fig. 1.8 Some well-known variable density undersampling patterns. **a** The variable density radial undersampling pattern, **b** the variable density random undersampling pattern based on the estimated probability density function, **c** the variable density random undersampling pattern based on the Poisson distribution

sition to prevent motion artifacts. We have seen that CS improves the speed of data acquisition process. The CS alone or the combination of the CS with PI in clinical MRI can greatly improve pediatric imaging [23].

Practical undersampling pattern must obey hardware (like gradient amplitude variation and slew rate) as well as physiological (nerve stimulation) limitations . A practical undersampling pattern should be either smoothly varying lines or curves to prevent frequent variation of the required gradients [15]. Some of the variable density sampling patterns for random undersampling in k-space are as follows.

Single-slice 2D MRI

In 2D MRI, readout direction, i.e., the frequency encode (k_x) direction is fully sampled using high-speed A/D converters. Here, the scan time is directly proportional to the total number of phase encode (k_y) lines. Thus, undersampling is required only in the phase encode (k_y) direction which may be carried out by randomly dropping some lines (as shown in Fig. 1.9a) in that direction. Generally, since clinical MRI uses Cartesian sampling, so with a little modification in the existing pulse sequence, one can implement 1D random undersampling pattern for CS-MRI [14].

Multi-slice 2D MRI

In multi-slice 2D MRI, random variable density undersampling is done only in the phase encode (k_y) direction for different slices, i.e., in the k_y -z plane as shown in Fig. 1.9c) [14]. For example, brain MRI is performed using multi-slice 2D cartesian acquisitions.

The above two techniques of random undersampling are not very effective for CS-MRI as undersampling is carried out in the k_y direction of the k-space only.

3D MRI

In 3D MRI, a volume of data is acquired using three gradients, i.e., two phase encode gradients, in phase encode (k_y) and slice encode (k_z) directions, respectively and one frequency encode gradient, in the frequency encode direction (k_x). Therefore, we can effectively undersample the k-space in both the k_y and k_z directions which results in high 2D incoherence. Figure 1.9e shows random undersampling of 3D MRI. The total scan time in 3D MRI is directly proportional to the rate of undersampling in k_y and k_z directions [14].

Lustig et al. [11], proposed Poisson disc undersampling pattern with 2D acceleration in auto-calibrating parallel MRI (PMRI). Vasanawala et al. [23], proposed an optimal variable density Poisson disc undersampling pattern for practical CS-MRI. In PMRI, the data are correlated in multiple coils, so close sampling in k-space is inefficient. In pure random undersampling, there is tendency of getting either very close samples which is wasteful or large gap samples which causes loss of information. Besides, it is also having physiological limitation as mentioned earlier. On the other hand, Poisson disc undersampling overcomes these drawbacks as in Poisson disc sampling the points are tightly packed, randomly generated, and keep a minimum gap between any two neighboring points. For these reasons, variable density Poisson disc undersampling scheme may be a good choice for 3D clinical MRI.

Dynamic MRI

Dynamic MRI is a multidimensional signal acquisition with time as an extra dimension. In dynamic MRI, data acquisition is performed in spatial frequency-time, i.e., $(k - t)$ space. To improve the incoherence, random undersampling is performed in $(k - t)$ space instead of uniform sampling [15].

1.8 Some Implementations of CS-MRI for Clinical Applications

Dr. Shreyas Vasanawala, a pediatric radiologist at Lucile Packard Children's hospital is the pioneer in making the compressed sensing technology a reality in clinical MRI. He along with his research group successfully integrated the compressed sensing based clinical MR image reconstruction setup with an existing scanner for pediatric body imaging. They compressively acquired the k-space data and then applied efficient recovery algorithms to reconstruct high-resolution 2D MR images within a couple of minutes. Further, they also extended their work to 3D MRI reconstruction.

In [23], they share their 2 years of research experience on the development of compressed sensing for clinical MRI. In some clinical applications, like, abdomen MRI required rapid imaging to avoid degradation of image quality due to organ motion. For this, the patient needs to hold the breath multiple times during the imaging process. Before the development of compressed sensing, a routine breath-hold MRI would require to hold breath 20–30 s depending on the field of view to acquire images at diagnostic resolution. Many children and old age patients, especially those having diseases of the heart and lungs cannot hold breath for such a long period.

Vasanawala's group were able to achieve twofold acceleration in clinical MRI using the compressed sensing. For example, using software–hardware integrated

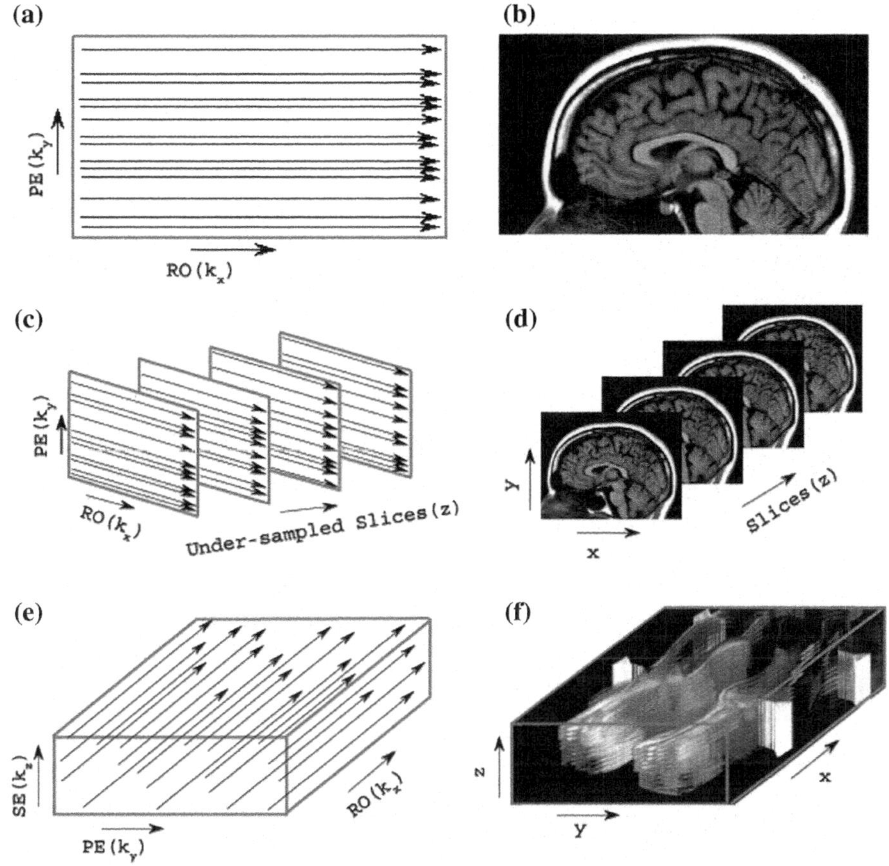

Fig. 1.9 Different types of Cartesian sampling patterns and MR images. **a** Single-slice 2D k-space under sampling pattern, **b** Single-slice 2D MR image, **c** multi-slice 2D k-space undersampling pattern, **d** multi-slice 2D MR images, **e** 3D k-space undersampling pattern and **f** 3D MR image

workstation, they were able to acquire diagnostic quality breath-hold abdomen images in just 10–15 s, which is relatively more comfortable for pediatric and aged patients [12, 18, 19]. In their work [23], they implemented a relatively simpler algorithm, namely, the projection over convex sets (POCS) using both Nvidia's Cuda and OpenMP on a dual socket six-core 2.67 GHz Intel Westmere system having four Nvidia's Tesla C1060. They reconstructed 3D MR volume first by decoupling each 3D problem into several independent 2D problems and then multiple 2D problems are solved in parallel using OpenMP and Cuda implementations. Again within each 2D problem, vector-paralalia idea is implemented for operations like, forward/ backward Fourier and wavelet transforms, interpolations, thresholding, etc., for efficient utilization of memory and bandwidth. Parallel implementation of 3D reconstruction was able to reconstruct high-resolution 3D MR volume in clinically feasible

Table 1.3 Required reconstruction time of POCS algorithm

Case	Data size	Reconstruction time (s)
1	192 × 320 × 66 × 8	24.4
2	320 × 206 × 108 × 12	59.3
3	192 × 320 × 110 × 32	120.7
4	172 × 230 × 188 × 32	169.9

time. Average reconstruction time of their implementations is shown in Table 1.3. The study was conducted as a part of their clinical routine checkups on 34 pediatric patients.

In clinical MRI practice, the cardiac imaging is the most challenging task. But, it provides most valuable information about complex anatomical structure. In conventional cardiac magnetic resonance (CMR), multiple breath-hold 2D cardiac cross section slices are acquired for the analysis of cardiac disorder. Recently, Dr. Vasanawala group proposed the idea of the 4D flow CMR [6–8, 20, 21]. It can analyze blood flow, contraction of cardiac chambers, etc., in any plane instead of a selected plan as in the conventional CMR which helps to detect anatomical or functional disorder through the cardiac cycle. It offers anatomical, functional, and flow information in a single free breathing acquisition of less then 10 min. For example, in [7], first gadolinium-based contrast agent was injected and then data were acquired using variable density Poisson disc undersampling with 50% acceleration on both phase and slice select directions. Finally, reconstructions were performed using the compressed sensing based parallel MRI reconstruction algorithm, ESPIRiT on a dedicated server equipped with high-frequency Intel Xeon E5-2670 processors and high performance NVIDIA graphics processing units.

The 4D flow CMR is a very rapidly growing cardiac imaging technique. It is expected that if 4D flow CMR is successfully implemented then it would definitely replace the conventional 2D CMR imaging [20]. In [6], 4D flow CMR was conducted on 54 patients for evaluation. In another recent work [21], authors integrated cloud-based compressed sensing reconstruction system for reconstruction of high-resolution 4D flow CMR from highly undersampled data. Although 4D flow CMR has several advantages, there are some drawbacks in clinical implementation like large amount of data, computational complexity of reconstruction techniques, and requirements of dedicated software- and hardware-integrated workstation for post-processing.

In addition to the above hardware-based approaches of clinical CS-MRI implementation, authors in [27] demonstrate a software-based approach for reconstruction of free breathing pediatric 3D abdomen MR angiography. Compressed sensing based parallel imaging technique is used for data acquisition with a variable density cartesian radial view undersampling scheme which also reduces motion artifacts. They did not use dedicated parallel processing hardware instead high-performance computing was used for compressed sensing based reconstruction. Average reconstruction time of their technique is approximately 5 min on a dedicated workstation having two Intel Xeon 2.3GHz 12 core processor with 256 GB RAM.

1.9 Conclusions

In this chapter, we study the basics of magnetic resonance imaging and the background of compressed sensing. We also observed how MRI is naturally compatible for the application of compressed sensing. In clinical MRI, compressed sensing is a highly potential alternative and practically viable tool for the improvement of the data acquisition time. Combination of compressed sensing with parallel imaging has the ability to reduce the need, duration, and strength of anaesthesia which would greatly improve the patient comfort and make MRI the most preferred diagnostic imaging tool with the state-of-the-art imaging technology.

References

1. Bushberg, J.T., Seibert, A.J., Leidholdt, E.M., Boone, J.M.: The Essential Physics of Medical Imaging. Lippincott Williams and Wilkins, PA (2012)
2. Candes, E.J., Romberg, J.K.: Signal recovery from random projections. In: Proceedings of SPIE Computational Imaging III, vol. 5674, pp. 76-86. San Jose (2005)
3. Candes, E., Wakin, M.: An introduction to compressive sampling. IEEE Signal Process. Mag. **25**(2), 21–30 (2008)
4. Candes, E.J., Romberg, J.K., Tao, T.: Robust uncertainty principles: exact signal reconstruction from highly incomplete frequency information. IEEE Trans. Inf. Theory **52**(2), 489–509 (2006)
5. Candes, E.J., Romberg, J.K., Tao, T.: Stable signal recovery from incomplete and inaccurate measurements. Commun. Pure Appl. Math. **59**(8), 1207–1223 (2006)
6. Chelu, R.G., van den Bosch, A.E., van Kranenburg, M., Hsiao, A., van den Hoven, A.T., Ouhlous, M., Budde, R.P.J., Beniest, K.M., Swart, L.E., Coenen, A., Lubbers, M.M., Wielopolski, P.A., Vasanawala, S.S., Roos-Hesselink, J.W., Nieman, K.: Qualitative grading of aortic regurgitation: a pilot study comparing CMR 4D flow and echocardiography. Int. J. Cardiovasc. Imaging **32**, 301–307 (2016)
7. Chelu, R.G., Wanambiro, K.W., Hsiao, A., Swart, L.E., Voogd, T., van den Hoven, A.T., van Kranenburg, M., Coenen, A., Boccalini, S., Wielopolski, P.A., Vogel, M.W., Krestin, G.P., Vasanawala, S.S., Budde, R.P., Roos-Hesselink, J.W., Nieman, K.: Cloud-processed 4D CMR flow imaging for pulmonary flow quantification European. J. Radiol. **85**, 1849–1856 (2016)
8. Cheng, J.Y., Zhang, T., Pauly, J.M., Vasanawala, S.S.: Feasibility of ultra-high-dimensional flow imaging for rapid pediatric cardiopulmonary MRI. J. Cardiovasc. Magn. Reson. **18**(Suppl 1), 217 (2016)
9. Deka, B., Datta, S.: A practical under-sampling pattern for compressed sensing MRI. In: Advances in Communication and Computing. Lecture Notes in Electrical Engineering, vol. 347, chap. 9, pp. 115–125. Springer, India (2015)
10. Hsiao, A., Lustig, M., Alley, M.T., Murphy, M.J., Vasanawala, S.S.: Evaluation of valvular insufficiency and shunts with parallel-imaging compressed-sensing 4D phase-contrast MR imaging with stereoscopic 3D velocity-fusion volume-rendered visualization. Radiology **265**, 87–95 (2012)
11. Lustig, M., Alley, M.T., Vasanawala, S., Donoho, D., Pauly, J.: L_1-SPIRiT: autocalibrating parallel imaging compressed sensing. In: 17th Annual Meeting of ISMRM, p. 379. Honolulu, Hawaii (2009)
12. Lustig, M., Keutzer, K., Vasanawala, S.: Introduction to parallelizing compressed sensing magnetic resonance imaging. In: Patterson, D., Gannon, D., Wrinn, M. (eds.) The Berkeley Par Lab: Progress in the Parallel Computing Landscape, pp. 105–139. Microsoft Corporation (2013)

13. Lustig, M., Pauly, J.: SPIRiT: iterative self-consistent parallel imaging reconstruction from arbitrary k-space. Magn. Reson. Med. **64**(2), 457–471 (2010)
14. Lustig, M., Donoho, D., Pauly, J.M.: Sparse MRI: the application of compressed sensing for rapid MR imaging. Magn. Reson. Med. **58**, 1182–1195 (2007)
15. Lustig, M., Donoho, D., Santos, J., Pauly, J.: Compressed sensing MRI. IEEE Signal Process. Mag. **25**(2), 72–82 (2008)
16. Majumdar, A.: Compressed Sensing for Magnetic Resonance Image Reconstruction. Cambridge University Press, New York (2015)
17. McRobbie, D.W., Moore, E.A., Graves, M.J., Prince, M.R.: MRI from Picture to Proton, 2nd edn. Cambridge University Press, Cambridge (2006)
18. Murphey, M., Keutzer, K., Vasanawala, S., Lustig, M.: Clinically feasible reconstruction time for L_1-SPIRiT parallel imaging and compressed sensing MRI. In: Proceedings of the International Society for Magnetic Resonance in Medicine, pp. 48–54 (2010)
19. Murphey, M., Alley, M., Demmel, J., Keutzer, K., Vasanawala, S., Lustig, M.: Fast L_1-SPIRiT compressed sensing parralel imaging MRI: scalable parallel implementation and clinically feasible runtime. IEEE Trans. Med. Imaging **31**(6), 1250–1262 (2012)
20. Saru, R.G., Wanambiro, K., Hsiao, A., Boccalini, S., Coenen, A., Budde, R., Wielopolski, P., Vasanawala, S., Roos-Hesselink, J., Nieman, K.: Global left ventricular function quantification with CMR 4D Flow. J. Cardiovasc. Magn. Reson. **18**, 308 (2016)
21. Saru, R.G., Wanambiro, K., Hsiao, A., Swart, L.E., Boccalini, S., Vogel, M., Budde, R., Vasanawala, S., Roos-Hesselink, J., Nieman, K.: Remote CMR 4D flow quantification of pulmonary flow. J. Cardiovasc. Magn. Reson. **18**, 307 (2016)
22. Usman, M., Batchelor, P.G.: Optimized Sampling Patterns for Practical Compressed MRI. Marseille, France (2009)
23. Vasanawala, S., Murphy, M., Alley, M., Lai, P., Keutzer, K., Pauly, J., Lustig, M.: Practical parallel imaging compressed sensing MRI: summary of two years of experience in accelerating body MRI of pediatric patients. In: IEEE International Symposium on Biomedical Imaging: From Nano to Macro 2011, pp. 1039–1043. Chicago, IL (2011)
24. Vasanawala, S.S., Lustig, M.: Advances in pediatric body MRI. Pediatr Radiol. **41**(Suppl 2), S549–S554 (2011)
25. Vasanawala, S., Alley, M., Hargreaves, B., Barth, R., Pauly, J., Lustig, M.: Improved pediatric MR imaging with compressed sensing. Radiology **256**(2), 607–616 (2010)
26. Yang, J., Zhang, Y., Yin, W.: A fast alternating direction method for TVL1-L2 signal reconstruction from partial Fourier data. IEEE J. Sel. Top. Signal Process. **4**(2), 288–297 (2010)
27. Zhang, T., Yousaf, U., Hsiao, A., Cheng, J.Y., Alley, M.T., Lustig, M., Pauly, J.M., Vasanawala, S.S.: Clinical performance of a free-breathing spatiotemporally accelerated 3-D time-resolved contrast-enhanced pediatric abdominal MR angiography. Pediatr Radiol. **45**(11), 1635–1643 (2015)

Chapter 2
CS-MRI Reconstruction Problem

Abstract Compressed sensing MRI (CS-MRI) seeks good quality MR image reconstruction from relatively less number of measurements than the traditional Nyquist sampling theorem. This in return increases the computational effort for reconstruction which may be dealt with some efficient solvers based on convex optimization. To reconstruct MR image from undersampled Fourier data, an underdetermined system of equations is needed to be solved with some additional information as regularization priors, like, compressibility of MR images in the spatial as well as wavelet domains.

2.1 Introduction

Compressed sensing (CS) has drawn a lot of interests from the signal processing community in the last one decade or so. Theoretically, it implies that a sparse signal $\mathbf{x} \in \mathbb{R}^n$ is to be acquired using a sensing/ measurement matrix $\mathbf{A} \in \mathbb{R}^{m \times n}$ with $m << n$ so that the measured data $\mathbf{y} = \mathbf{Ax}$. Now, $\mathbf{x}$ can be exactly reconstructed from $\mathbf{y} \in \mathbb{R}^m$ if both $\mathbf{x}$ and $\mathbf{A}$ satisfy the requirements of CS theory as discussed in Chap. 1. As the system is underdetermined, the conventional approach of reconstruction of $\mathbf{x}$ from the given $\mathbf{y}$ and $\mathbf{A}$ is to solve a least squares problem which generally yields a dense solution. If we assume that $\mathbf{x}$ is sufficiently sparse in the acquisition domain itself, and the columns of $\mathbf{A}$ are mutually orthogonal then one can exactly reconstruct $\mathbf{x}$ by solving the following minimization problem:

$$\begin{aligned} &\min_{\mathbf{x}} \ \|\mathbf{x}\|_0 \\ &\text{subject to } \ \mathbf{y} = \mathbf{Ax} \end{aligned} \tag{2.1}$$

where $\|\mathbf{x}\|_0$ indicates total number of nonzero coefficients in $\mathbf{x}$. The problem defined in Eq. 2.1 is non-convex in nature and solving $\|\mathbf{x}\|_0$-minimization is computationally prohibitive. A common way of replacement of $\|\mathbf{x}\|_0$ is the ℓ_1-minimization of $\mathbf{x}$, i.e., $\|\mathbf{x}\|_1$-minimization, which can be stated as

B. Deka and S. Datta, *Compressed Sensing Magnetic Resonance Image Reconstruction Algorithms*, Springer Series on Bio- and Neurosystems 9,
https://doi.org/10.1007/978-981-13-3597-6_2

$$\min_{\mathbf{x}} \ \|\mathbf{x}\|_1 \quad \text{subject to} \ \mathbf{y} = \mathbf{A}\mathbf{x} \tag{2.2}$$

where $\|\mathbf{a}\|_1 = \sum_{i=1}^{n} |a_i|$. The above minimization problem is convex and computationally tractable which can be solved using conventional linear programming approach or primal-dual interior-point method. But, computational costs of these algorithms are quite high and sometimes it becomes impractical for large-scale problems [2, 27].

Besides stability and scalability one more important requirements for practical application is the robustness to noise, i.e., measured data $\mathbf{y}$ may also contain noise from surrounding environment. By considering the noise during measurement, we can relax the equality constraint and rewrite the problem in Eq. 2.2 as

$$\min_{\mathbf{x}} \|\mathbf{x}\|_1 \quad \text{subject to} \ \|\mathbf{y} - \mathbf{A}\mathbf{x}\|_2 \leqslant \varepsilon, \tag{2.3}$$

where ε is a positive constant indicating the nose level. The above problem is well known in the literature and given the name basis pursuit denoising (BPDN) problem [1, 3].

An alternative approach to solve the problem given in Eq. 2.2 is by solving greedy algorithms, like, the orthogonal matching pursuit (OMP) [4, 20], the least angle regression (LARS) [7], etc. They are extremely good when the signal is sufficiently sparse. But performance is degraded when signal sparsity is reduced. Moreover, there is no theoretical guarantee of them converging to the global solution.

Greedy algorithms are simpler, fast, and suitable for hardware implementation. If we somehow identify the support of sparse signal then reconstruction is quite simple, i.e., by considering the basis functions corresponding to those positions one can get good quality of reconstruction. Unfortunately, we do not have any *a priori* information about the signal support. Greedy algorithms find the support information iteratively. Based on the support selection approach, a number of greedy algorithms exist [13, Chapter 1, p. 38]. Among them the simplest one is the matching pursuit(MP) [15] which is summarized in Algorithm 1. In the first step, it finds residue $\mathbf{r} = \mathbf{y} - \mathbf{A}\mathbf{x}^k$ which is to be initialized as $\mathbf{r} = \mathbf{y}$. Next, the column of $\mathbf{A}$ which gives maximum correlation with the residue is selected and then projection of the column on to the current residue is obtained and added to the previous solution $\mathbf{x}^k$ to get the current solution $\mathbf{x}^{k+1}$. Subsequently, a new residue is obtained by subtracting the product $\mathbf{A}\mathbf{x}^{k+1}$ from $\mathbf{y}$. The above three steps are repeated until convergence. Generally, there are two stoping criteria for these algorithms: (a) algorithmic steps are repeated until k-support information has been found. In this case it is assumed that signal is k-sparse. But in practice, it is very difficult to predict the exact value of k. (b) algorithmic steps are repeated until residue is below a predefined value, i.e., $\|\mathbf{y} - \mathbf{A}\mathbf{x}\|_2^2 \leqslant \varepsilon$.

Algorithm 1 Matching Pursuit(MP) Algorithm

Input: $\mathbf{A}, \mathbf{y}$
Initialization: $\Lambda \leftarrow \{\ \}, \ \mathbf{x}^0 \leftarrow \mathbf{0}, \ k \leftarrow 0$
1: **while** not converge **do**
2: $\quad \mathbf{r} \leftarrow (\mathbf{y} - \mathbf{A}\mathbf{x}^k)$
3: $\quad \lambda^k \leftarrow \arg\max_j \left|\mathbf{A}_j^T \mathbf{r}\right|$
4: $\quad \Lambda \leftarrow \Lambda \cup \lambda^k$
5: $\quad \mathbf{x}^{k+1} \leftarrow \mathbf{x}^k + \mathbf{A}_{\lambda^k}^T \mathbf{r}$
6: $\quad k \leftarrow k+1$
7: **end while**
Output: $\hat{\mathbf{x}} \leftarrow \mathbf{x}^k$

There is no theoretical guarantee that the MP can successfully recover sparse signals from an underdetermined system. It is commonly used for tight frame or orthogonal systems. Another disadvantage in MP is that there is no assurance that same index will not appear more than once. To overcome this drawback, the orthogonal matching pursuit (OMP) was introduced [20]. It provides the guarantee that the same atom will not be selected more than once. OMP steps are summarized in Algorithm 2. At each iteration solution, $\mathbf{x}$ is obtained by solving a least squares problem. It also assures that estimated $\mathbf{x}^{k+1}$ at each iteration is orthogonal to the residue $\mathbf{y} - \mathbf{A}\mathbf{x}^{k+1}$. However, the computational cost of the OMP is slightly more than that of the MP.

Algorithm 2 Orthogonal Matching Pursuit(OMP) Algorithm

Input: $\mathbf{A}, \mathbf{y}$
Initialization: $\Lambda \leftarrow \{\ \}, \ \mathbf{x}^0 \leftarrow \mathbf{0}, \ k \leftarrow 0$
1: **while** not converge **do**
2: $\quad \mathbf{r} \leftarrow (\mathbf{y} - \mathbf{A}\mathbf{x}^k)$
3: $\quad \lambda^k \leftarrow \arg\max_j \left|\mathbf{A}_j^T \mathbf{r}\right|$
4: $\quad \Lambda \leftarrow \Lambda \cup \lambda^k$
5: $\quad \mathbf{x}^{k+1} \leftarrow \min_{\mathbf{x}} \|\mathbf{y} - \mathbf{A}_\Lambda \mathbf{x}_\Lambda\|_2^2$
6: $\quad k \leftarrow k+1$
7: **end while**
Output: $\hat{\mathbf{x}} \leftarrow \mathbf{x}^k$

One common disadvantage in both the MP and the OMP is that they obtain one support index at each iteration. To overcome this problem, the concept of soft thresholding is combined with the OMP. Here, the coefficient whose magnitude is greater than a predefined threshold value corresponding indices are considered as support indices. This greedy method is known as the stagewise OMP (StOMP) [6]. In the StOMP, solution $\mathbf{x}^{k+1}$ at each iteration is obtained in the same way as in case of the OMP, i.e., by solving a least squares problem with available support information.

There are some other well-known greedy algorithms, namely, Compressive Sampling Matching Pursuit (CoSamp) [18], Block OMP (BOMP) [8], Group OMP [14], Generalized OMP (gOMP) [26], Regularized OMP (ROMP) [19], and Simultaneous OMP (SOMP) [22]. Although a few works demonstrate [9, 23] the application of OMP or its variants for CS-MRI reconstruction of dynamic MRI images but their performances are not comparable to those obtained using the basis pursuit technique due to the fact that the later also guarantees theoretical convergence for large scale settings [10] or when the images have relatively poor SNR. Thus, we focus on to the basis pursuit approach for solving the highly nonlinear problem of CS-MRI reconstruction.

2.2 CS-MRI Problem Formulation

MR signals carry information of spatially varying resonant frequencies from a part of the human body under investigation. MRI images or higher dimensional data in spatial domain are obtained by taking the inverse Fourier transform of the scanned data. Like natural images, MR images are sufficiently compressible in a transform domain, e.g., in the wavelet domain. Now, with this a priori knowledge as a penalty one can formulate a BPDN problem for CS-MRI reconstruction. Let us consider an image $\mathbf{x}$ which is compressible over a transform Ψ, i.e., $\mathbf{x} = \Psi^T \boldsymbol{\alpha}$, where $\boldsymbol{\alpha}$ represents a sparse vector. If $\mathbf{y}$ denotes the undersampled data in k-space, then we can reconstruct $\mathbf{x}$ by solving the following constrained optimization problem:

$$\begin{aligned} &\min || \Psi \mathbf{x} ||_1 \\ &\text{subject to } ||\mathbf{F}_u \mathbf{x} - \mathbf{y}||_2 \leq \varepsilon \end{aligned} \tag{2.4}$$

where $\mathbf{F}_u$ is the partial Fourier transform operator constructed by the k-space undersampling scheme as mentioned in Chap. 1 and ε is the root mean-squared error.

The equivalent unconstrained problem, i.e., the Lagrangian form is given by

$$(P_1): \quad \hat{\mathbf{x}} = \underset{\mathbf{x}}{\text{argmin}} \; ||\mathbf{F}_u \mathbf{x} - \mathbf{y}||_2^2 + \lambda || \Psi \mathbf{x} ||_1 \tag{2.5}$$

where λ is a regularization parameter to balance between the data consistency and the sparsity terms. The solution of P_1 gives a good approximation of the original image. Typically, MR images are compressible in wavelets [11, 28], therefore, Ψ may be formed by properly chosen wavelet bases.

In addition to the sparsification of the image in the wavelet domain, we can add another very useful sparsifying norm to increase the overall sparsity. The total variation (TV) norm [21], i.e., sum of absolute variations of pixel intensity or the gradient sparsity of an MR image is commonly used as MR images are not only smooth but contains many texture regions as well in it. Therefore, ℓ_1 combined with TV would improve the reconstruction result by enhancing the sparsity of the image in the transform as well as in the spatial domain. Thus, we rewrite the P_1 problem for CS-MRI reconstruction as

$$(P_2): \quad \hat{\mathbf{x}} = \underset{\mathbf{x}}{\operatorname{argmin}} \, ||\mathbf{F}_u\mathbf{x} - \mathbf{y}||_2^2 + \lambda_1||\Psi\mathbf{x}||_1 + \lambda_2\,||\mathbf{x}||_{TV} \tag{2.6}$$

where λ_1 and λ_2 are regularization parameters that establish trade off between the data consistency and the sparsity, the term $||\mathbf{x}||_{TV}$ is the total variation of $\mathbf{x}$ in isotropic sense, i.e.,
$||\mathbf{x}||_{TV} = \sum_i \sqrt{\{(\nabla_h\mathbf{x})_i\}^2 + \{(\nabla_v\mathbf{x})_i\}^2}$, ∇_h and ∇_v being corresponding first-order horizontal and vertical difference operators.

P_2, due to the non-smoothness of both ℓ_1 and TV regularization terms, is not differentiable. Lustig et al. [11] solved this problem using the nonlinear conjugate gradient method but the entire process is relatively slower for implementation of the CS-MRI. To overcome this problem, Ma et al. [12] and Yang et al. [28] solved the above problem using operator and variable splitting techniques, respectively, discussed in the next chapter. Recently, Huang et al. [10] and authors in [5] solved it using a hybrid splitting technique where a combination of both operator and variable splitting techniques is used.

These techniques have been successful in significantly reducing the reconstruction time compared to [11]. The results exhibit almost no visual loss of information even at 20% sampling ratio. However, at lower sampling ratios for example at 15%, although visual information are still preserved, artifacts would coexist with the useful visual information. So, we can conclude that at least 20% sampling ratio is required for better reconstruction. Recently, some algorithms are able to produce good quality reconstructed images within a few seconds. For example, to reconstruct a 256×256 brain MR image, the Fast Composite Splitting Algorithm (FCSA) [10] requires 4–5 s in MATLAB on a 3.4 GHz PC having 32-bit OS and 2GB RAM to achieve an average PSNR of 31–35 dB.

For clinical implementation of compressed sensing MRI, reconstruction time is a barrier [25]. We need to reconstruct a large number of 2D MRI slices within a couple of minutes, i.e., we need to solve each 2D problem in approximately half a second time without sacrificing the quality of reconstruction. This is a challenging task because each problem contains computationally expensive operations like forward and backward Fourier transforms, sparsifying transforms with wavelets, and then for parallel MRI, multiplication with coil sensitivity profiles or convolution operations. Above all, these nonlinear optimization algorithms are to be solved iteratively which would aggravate the problem further.

Dr. Shreyas Vasanawala's group in 2010 for the first time implemented compressed sensing MRI technology in clinical setting at the Lucile Packard Children's Hospital Stanford. They used 3D spoiled-gradient-echo with variable density Poisson disk undersampling pattern to accelerate data acquisition. Reconstruction was performed by the Projection Over Convex Set algorithm implemented using parallel architectures in multicore CPU and General Purpose Graphics Processors (GPGPU) [16, 17, 24, 25].

In the next chapter, we will discuss some of the very popular fast compressed sensing MR image reconstruction algorithms with their mathematical details.

2.3 Conclusions

This chapter demonstrates a brief mathematical background of compressed sensing in a more general sense for an arbitrary sparse signal and the CS-MRI problem in particular. The main focus of this chapter is the formulation of CS-MRI problem in a convex optimization framework due its advantages. It also briefly discusses about the present state-of-the-art for CS-MRI reconstruction in a clinical setting and its related hardware.

References

1. Candes, E.J., Romberg, J.K., Tao, T.: Stable signal recovery from incomplete and inaccurate measurements. Commun. Pure Appl. Math. **59**(8), 1207–1223 (2006)
2. Candes, E., Wakin, M., Boyd, S.: Enhancing sparsity by reweighted L1 minimization. J. Fourier Anal. Appl. **14**(5), 877–905 (2008)
3. Chen, S.S., Donoho, D.L., Saunders, M.A.: Atomic decomposition by basis pursuit. SIAM J. Sci. Comput. **20**, 33–61 (1998)
4. Davis, G., Mallat, S., Avellaneda, M.: Adaptive greedy approximations. Constr. Approx. **13**(1), 57–98 (1997)
5. Deka, B., Datta, S.: High throughput MR image reconstruction using compressed sensing. In: Proceedings of the 2014 Indian Conference on Computer Vision Graphics and Image Processing, ICVGIP14, pp. 89:1–89:6. ACM, Bangalore, India (2014)
6. Donoho, D.L., Tsaig, Y., Drori, I., Starck, J.L.: Sparse solution of underdetermined systems of linear equations by stagewise orthogonal matching pursuit. IEEE Trans. Inf. Theory **58**(2), 1094–1121 (2012)
7. Efron, B., Hastie, T., Johnstone, I., Tibshirani, R.: Least angle regression. Ann. Stat. **32**(2), 407–451 (2004)
8. Eldar, Y.C., Kuppinger, P., Bolcskei, H.: Block-sparse signals: uncertainty relations and efficient recovery. IEEE Trans. Signal Process. **58**(6), 3042–3054 (2010)
9. Gamper, U., Boesiger, P., Kozerke, S.: Compressed sensing in dynamic MRI. Magn. Reson. Med. **59**(2), 365–373 (2008)
10. Huang, J., Zhang, S., Metaxas, D.N.: Efficient MR image reconstruction for compressed MR imaging. Med. Image Anal. **15**(5), 670–679 (2011)
11. Lustig, M., Donoho, D., Pauly, J.M.: Sparse MRI: the application of compressed sensing for rapid MR imaging. Magn. Reson. Med. **58**, 1182–1195 (2007)
12. Ma, S., Yin, W., Zhang, Y., Chakraborty, A.: An efficient algorithm for compressed MR imaging using total variation and wavelets. In: IEEE Conference on Computer Vision and Pattern Recognition (CVPR 2008), pp. 1–8. Anchorage, AK (2008)
13. Majumdar, A.: Compressed Sensing for Magnetic Resonance Image Reconstruction. Cambridge University Press, Delhi (2015)
14. Majumdar, A., Ward, R.K.: Fast group sparse classification. Can. J. Electr. Comput. Eng. **34**(4), 136–144 (2009)
15. Mallat, S., Zhang, Z.: Matching pursuits with time-frequency dictionaries. IEEE Trans. Signal Process. **41**, 3397–3415 (1993)

16. Murphey, M., Keutzer, K., Vasanawala, S., Lustig, M.: Clinically feasible reconstruction time for L_1-SPIRiT parallel imaging and compressed sensing MRI. In: Proceedings of the International Society for Magnetic Resonance in Medicine, pp. 48–54 (2010)
17. Murphey, M., Alley, M., Demmel, J., Keutzer, K., Vasanawala, S., Lustig, M.: Fast L_1-SPIRiT compressed sensing parralel imaging MRI: scalable parallel implementation and clinically feasible runtime. IEEE Trans. Med. Imaging **31**(6), 1250–1262 (2012)
18. Needell, D., Tropp, J.: CoSaMP: iterative signal recovery from incomplete and inaccurate samples. Appl. Comput. Harmon. Anal. **26**(3), 301–321 (2009)
19. Needell, D., Vershynin, R.: Signal recovery from incomplete and inaccurate measurements via regularized orthogonal matching pursuit. IEEE J. Sel. Top. Signal Process. **4**(2), 310–316 (2010)
20. Pati, Y.C., Rezaiifar, R., Rezaiifar, Y.C.P.R., Krishnaprasad, P.S.: Orthogonal matching pursuit: recursive function approximation with applications to wavelet decomposition. In: Proceedings of the 27th Annual Asilomar Conference on Signals, Systems, and Computers, pp. 40–44 (1993)
21. Rudin, L.I., Osher, S., Fatemi, E.: Nonlinear total variation based noise removal algorithms. Phys. D **60**, 259–268 (1992)
22. Tropp, J.A., Gilbert, A.C., Strauss, M.J.: Algorithms for simultaneous sparse approximation. part i. Signal Process. **86**(3), 572–588 (2006)
23. Usman, M., Prieto, C., Odille, F., Atkinson, D., Schaeffter, T., Batchelor, P.G.: A computationally efficient OMP-based compressed sensing reconstruction for dynamic MRI. Phys. Med. Biol. **56**(7), N99–N114 (2011)
24. Vasanawala, S., Murphy, M., Alley, M., Lai, P., Keutzer, K., Pauly, J., Lustig, M.: Practical parallel imaging compressed sensing MRI: summary of two years of experience in accelerating body MRI of pediatric patients. In: IEEE International Symposium on Biomedical Imaging: From Nano to Macro 2011, pp. 1039-1043. Chicago, IL (2011)
25. Vasanawala, S., Alley, M., Hargreaves, B., Barth, R., Pauly, J., Lustig, M.: Improved pediatric MR imaging with compressed sensing. Radiology **256**(2), 607–616 (2010)
26. Wang, J., Kwon, S., Shim, B.: Generalized orthogonal matching pursuit. IEEE Trans. Signal Process. **60**(12), 6202–6216 (2012)
27. Yang, A.Y., Ganesh, A., Zhou, Z., Sastry, S., Ma, Y.: A review of fast L_1-minimization algorithms for robust face recognition (2010) CoRR arXiv:abs/1007.3753
28. Yang, J., Zhang, Y., Yin, W.: A fast alternating direction method for TVL_1-L_2 signal reconstruction from partial Fourier data. IEEE J. Sel. Top. Signal Process. **4**(2), 288–297 (2010)

Chapter 3
Fast Algorithms for Compressed Sensing MRI Reconstruction

Abstract Extensive research work is being carried out in the area of fast convex optimization-based compressed sensing magnetic resonance (MR) image reconstruction algorithms. The main focus here is to achieve throughputs of clinical compressed sensing MR image reconstruction in terms of quality of reconstruction and computational time. In this chapter, we briefly review some of the recently developed convex optimization-based algorithms for compressed sensing MR image reconstruction. All these algorithms may be classified broadly into four categories based on their approaches of solving the reconstruction/recovery problem. We then detail algorithms of each category with sufficient mathematical details and report their relative advantages and disadvantages.

3.1 Introduction

In CS-MRI, a straightforward linear reconstruction of MR images from the k-space data is not possible as they are randomly undersampled. Therefore, the reconstruction is basically an inverse ill-posed problem which may be solved either using greedy or optimization-based techniques. Greedy algorithms iteratively approximate supports and feature coefficients, until it meets a specific convergence criterion. There are a large number of greedy algorithms. General steps of a greedy algorithm are (a) selection of a set of basis functions for representation of the signal and then (b) update the coefficient vector. Mallat and Zhang have proposed a greedy algorithm known as the matching pursuit (MP) algorithm [45]. It iteratively selects a set of atoms from a dictionary for best sparse representation of the given signal. Among different greedy approaches, MP is the simplest one. Pati et al. proposed an extension of MP algorithm known as the orthogonal matching pursuit (OMP) algorithm [50]. The main difference is that in the OMP algorithm in each iteration, coefficients are updated by orthogonal projection of the signal on the selected atoms of the dictionary so that no atom is selected twice in subsequent iterations. Matching pursuit algorithm selects only one atom of the dictionary at each iteration and update the corresponding coefficient. At each iteration, it finds the best-correlated dictionary atom with the residual. Greedy algorithms are simple, fast, and easy to implement.

B. Deka and S. Datta, *Compressed Sensing Magnetic Resonance Image Reconstruction Algorithms*, Springer Series on Bio- and Neurosystems 9,
https://doi.org/10.1007/978-981-13-3597-6_3

At each iteration, they made a local optimum selection with the hope that it would lead to a global optimum. Matching pursuit algorithms are designed mainly for tight-frame or orthogonal systems. The main disadvantage associated with these algorithms is that there is no theoretical guarantee that MP algorithms will converge even after system satisfies the RIP condition [44]. Moreover, greedy algorithms required a large number of iterations to estimate the solution.

All greedy algorithms are based on the same philosophy, they start with a zero vector and then, estimate new nonzero components iteratively. They work well when the signal to be reconstructed is sufficiently sparse. Hence, these greedy techniques are not suitable for the arbitrary underdetermined system as in the case of CS-MRI. On the other hand, ℓ_1-norm minimization algorithm gives a well approximation of ground truth [56] within a reasonable number of iterations. Per-iteration computational complexity of the ℓ_1-norm minimization-based algorithms may be relatively more as compared to greedy algorithms but the total number iterations are less. Most importantly, the ℓ_1-norm minimization-based algorithms have strong theoretical evidence of convergence within a finite number of iterations. The real-world requirements for practical application are scalability, robustness, and global convergence of the algorithm. Because the amount of acquired data is not predefined as it varies with the particular application in hand and may also contain some amount of noise with it.

Due to these reasons, the ℓ_1-norm minimization-based convex optimization techniques are most popular and successful in CS-MRI. In this chapter, we give a comprehensive review of the recent developments of convex optimization-based CS-MRI reconstruction algorithm. Some of the well known algorithms are the *Projections Over Convex Set* (POCS) [41, 60], the *Interior-Point Method* [13, Chapter 11], the *Truncated Newton Interior-Point Method* (TNIPM) [40], the *Iterative Shrinkage-Thresholding* (IST) [21], the *Nonlinear Conjugate Gradient* (NCG) [42], the *Two-Step Iterative Shrinkage-Thresholding* (TwIST) [11], the *Gradient Projection for Sparse Reconstruction* (GPSR) [29], the *Spectral Projected Gradient algorithm-* SPGL1 [6], the *Fast Iterative Shrinkage-Thresholding Algorithm* (FISTA) [4], the *Sparse Reconstruction by Separable Approximation* (SpaRSA) [54], the *Split-Bregman method* [33], the *Split Augmented Lagrangian Shrinkage Algorithm* (SALSA) [1], the *Reconstruction from Partial Fourier data* (RecPF) [58], the *Nesterov's Algorithm-* NESTA [5], the *Alternating Direction Method* (ADM) [57], the *Composite Splitting Algorithm* (CSA) [38], the *Fast Composite Splitting Algorithm* (FCSA) [39], and the recently proposed high-throughput algorithm [22] using the combination of composite splitting denoising (CSD) and the ALM (ALM-CSD).

One of the most important goals of these algorithms is to decompose or split the given problem into smaller subproblems so that they can be solved in parallel to speed up the overall execution [25, Chapter 1]. In the following, we first give a very brief introduction on the formulation of each of the above algorithms with some details about their convergence. Experimental results are then summarized in the next chapter.

Introduction to Splitting: Projection onto the Convex Sets

The POCS is the simplest splitting algorithm based on alternating projections onto convex sets [15, 16, 41]. Rewriting the P_1 problem (see Chap. 2) of CS-MRI reconstruction to solve for $\boldsymbol{\alpha}$

$$\underset{\boldsymbol{\alpha}}{\operatorname{argmin}} \, ||\mathbf{F}_u \Psi^T \boldsymbol{\alpha} - \mathbf{y}||_2 + \lambda ||\boldsymbol{\alpha}||_1, \tag{3.1}$$

where $\boldsymbol{\alpha} = \Psi \mathbf{x}$ represent transform coefficients of the image. Since, $\boldsymbol{\alpha}$ is a solution of the above problem, the ℓ_1 ball $\mathbf{B} = \{||\boldsymbol{\alpha}||_1\}$ meets the hyperplane $\mathbf{H} = \{\mathbf{F}_u \Psi^T \boldsymbol{\alpha} - \mathbf{y}\}$ at the common point $\mathbf{B} \cap \mathbf{H} = \{\boldsymbol{\alpha}\}$. As both $\mathbf{B}$ and $\mathbf{H}$ are convex, we can apply the POCS algorithm of [16] to carry out projections first onto $\mathbf{H}$ and then onto $\mathbf{B}$, iteratively, to solve Eq. 3.1. The first projection estimates the measured data $\mathbf{y}$ from the estimated sparse coefficients and second projection estimates sparse coefficient vector $\boldsymbol{\alpha}$ by the simple soft-thresholding operation which is defined as

$$\text{soft}\,(a, b) = \begin{cases} \text{sgn}\,(a) \,.\, (|a| - b)\,, & \text{if } |a| > b \\ 0, & \text{otherwise} \end{cases} \tag{3.2}$$

The POCS algorithm is summarized in Algorithm 3, where $\mathbf{F}$ is the Fourier transform operator.

Algorithm 3 A Projection onto the Convex Sets (POCS) Algorithm

Input: $\mathbf{F}_u, \mathbf{y}, \Psi$
Initialization: $\mathbf{y}^{(0)} \leftarrow \mathbf{y},\ k \leftarrow 0$
while not converge **do**
  $\boldsymbol{\alpha}^{(k)} \leftarrow \Psi \mathbf{F}_u^T \mathbf{y}^{(k)}$
  $\boldsymbol{\alpha}^{(k)} \leftarrow soft\left(\boldsymbol{\alpha}^{(k)}, \lambda\right)$
  $\mathbf{y}^{(k+1)} \leftarrow \mathbf{F}_u \Psi^T \boldsymbol{\alpha}^{(k)}$
  $\mathbf{y}^{(k+1)}[i] \leftarrow \begin{cases} \mathbf{y}^{(k+1)}[i], & \text{if } \mathbf{y}[i] == 0 \\ \mathbf{y}[i], & \text{otherwise} \end{cases}$
  $k \leftarrow k + 1$
end while
Output: $\hat{\mathbf{x}} \leftarrow \Psi^T \boldsymbol{\alpha}$

Reconstructions algorithms involving other complex splitting techniques may be further classified into three broad categories, namely, the *operator splitting*, the *variable splitting*, and the *composite splitting*. Figure 3.1 shows the detailed classification of the existing state-of-the-art MR image reconstruction algorithms based on different splitting techniques. Algorithms without using explicit splitting, include, the interior-point method [13, Chapter 11], the truncated Newton interior-point method (TNIPM), the nonlinear conjugate gradient projection (NCGP) method [42], etc., are also shown in the same figure. Now, we discuss the algorithms in each category with details.

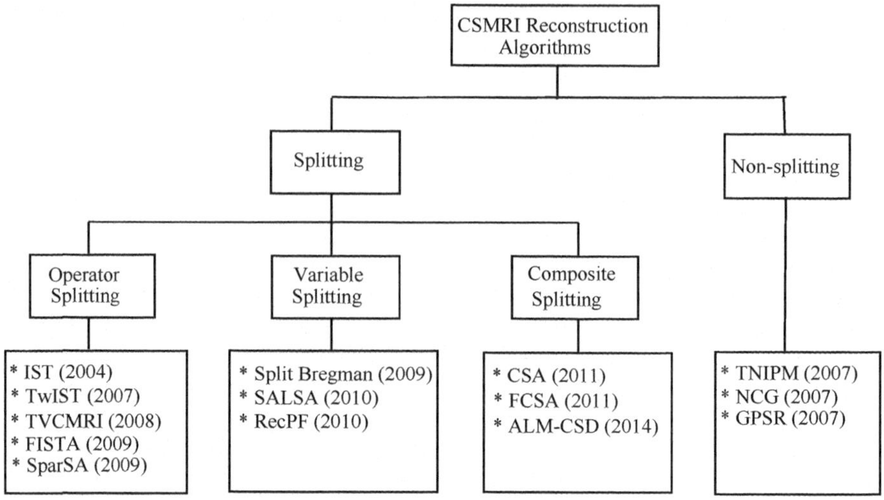

Fig. 3.1 Classification of MR image reconstruction algorithms based on the splitting technique

3.2 Operator Splitting Method

It is a popular technique in numerical linear algebra where two very closely related methods are the *explicit method* (forward step or gradient method) and the *implicit method* (backward step or the proximal point method) [25, Chapter 3, Sect. 2].

Consider the case of minimizing a convex function f which is defined and differentiable in everywhere. The differential equation

$$\frac{d}{dt}\mathbf{x}(t) = -\nabla f(\mathbf{x}(t)) \quad . \tag{3.3}$$

is known as the gradient flow for f [49, Chapter 4]. The points of the gradient flow in equilibrium are the exact minimizers of f.

The discretization of Eq. 3.3 with step size $\lambda > 0$ leads to

$$\frac{\mathbf{x}^{(k+1)} - \mathbf{x}^{(k)}}{\lambda} = -\nabla f\left(\mathbf{x}^{(k)}\right) \tag{3.4}$$

which yields $\mathbf{x}^{(k+1)}$ as-

$$\mathbf{x}^{(k+1)} = \mathbf{x}^{(k)} - \lambda \nabla f(\mathbf{x}^{(k)}). \tag{3.5}$$

The above formulation is called the forward Euler discretization. Since the forward step or the explicit method is similar to the steepest descent method, so convergence of this method depends on the proper selection of the step size λ. In order to get rid of the ill-conditioning of the forward step method, an alternative is the backward Euler approximation which may be done by a slight change of the above equation, i.e., by writing

$$\frac{\mathbf{x}^{(k+1)} - \mathbf{x}^{(k)}}{\lambda} = -\nabla f\left(\mathbf{x}^{(k+1)}\right). \tag{3.6}$$

We observe that $\mathbf{x}^{(k+1)}$ cannot be written explicitly in terms of $\mathbf{x}^{(k)}$, like the forward Euler method. For this reason, it is also named as the implicit method. Here, $\mathbf{x}^{(k+1)}$ is obtained by solving

$$\mathbf{x}^{(k+1)} = (I + \lambda_k \nabla f)^{-1} \mathbf{x}^{(k)}. \tag{3.7}$$

In general, if we replace ∇f by any arbitrary maximal monotone operator F, then Eq. 3.7 can be represented as

$$\mathbf{x}^{(k+1)} \in (I + \lambda_k F)^{-1} \mathbf{x}^{(k)}, \tag{3.8}$$

where $\{\lambda_k\}$ is a sequence of positive real numbers. This is known as the *proximal point method* [25]. The difficulty associated with the proximal point algorithm is due to the inverse operation $(I + \lambda F)^{-1}$.

We can split the operator F into two maximal monotone operators A and B such that $F = A + B$ and $(I + \lambda A)$ and $(I + \lambda B)$ are easily inverted. In general, the operator splitting technique is defined for a maximal monotone operator F that attempts to solve $0 \in F(\mathbf{x})$ by repeatedly applying operators of the form $(I + \lambda A)^{-1}$ and $(I + \lambda B)^{-1}$ [25, p. 63].

In the following, we discuss a few selected techniques in this category that are currently used for CS-MRI.

Forward–Backward Operator Splitting Method

It is a well-known class of operator splitting scheme. Here, for a positive scalar sequence $\{\lambda_k\}$, $\mathbf{x}^{(k+1)}$ is defined as [25, Chapter 3, Theorem 3.12]

$$\mathbf{x}^{(k+1)} \in (I + \lambda_k B)^{-1}\left(\mathbf{x}^{(k)} - \lambda_k A\mathbf{x}^{(k)}\right) = P_C\left(\mathbf{x}^{(k)} - \lambda_k A\mathbf{x}^{(k)}\right) \qquad \forall k \geqslant 0. \tag{3.9}$$

Thus forward–backward method is the combination of the two basic schemes, namely, the backward step or proximal point method and the forward step or the gradient method which are discussed above. The first term $(I + \lambda_k B)^{-1}$ of Eq. 3.9 is called the proximity operator [20]. It is the generalization of the notion of a projection operator onto a nonempty closed convex set. The projection $P_C\mathbf{x}$ of $\mathbf{x} \in \mathbb{R}^N$ onto the nonempty closed convex set $C \subset \mathbb{R}^N$ is the solution of

$$\underset{\mathbf{y} \in R^N}{\operatorname{argmin}} \tfrac{1}{2}||\mathbf{x} - \mathbf{y}||_2^2 + l_C(\mathbf{y}), \tag{3.10}$$

where $l_C(.)$ is the indicator function of C [19, Eq. 4]. In [46], Moreau replaced the indicator function by any arbitrary function φ. Thus, proximity operator of φ denoted by $prox_\varphi\mathbf{x}$ is the solution to the following minimization problem:

$$\underset{\mathbf{y}\in R^N}{\operatorname{argmin}} \tfrac{1}{2}||\mathbf{x}-\mathbf{y}||_2^2+\varphi(\mathbf{y}). \tag{3.11}$$

Consider a general optimization model as

$$\underset{\mathbf{x}}{\operatorname{argmin}} \{f(\mathbf{x})=h(\mathbf{x})+g(\mathbf{x})\} \tag{3.12}$$

where $h(\mathbf{x})$ and $g(\mathbf{x})$ are convex functions produced by splitting of $f(\mathbf{x})$, and $g(\mathbf{x})$ being non-differentiable in general. We can solve the above problem according to [20, Proposition 3.1, Eqs. 3.2–3.4]. This method is known as the forward–backward splitting process in the following optimization:

$$\mathbf{x}^{(k+1)}=prox_g\left(\mathbf{x}^{(k)}-\lambda_k\nabla h\left(\mathbf{x}^{(k)}\right)\right). \tag{3.13}$$

It consists of two separate steps, the first one is the forward step involving only h to compute $\mathbf{x}^{(k+\frac{1}{2})}=\mathbf{x}^{(k)}-\lambda_k\nabla h(\mathbf{x}^{(k)})$. This is followed by a backward step involving only g to compute $\mathbf{x}^{(k+1)}=prox_g(\mathbf{x}^{(k+\frac{1}{2})})$. For example, if $h(\mathbf{x})=\frac{1}{2}||\mathbf{A}\mathbf{x}-\mathbf{y}||_2^2$ and $g(\mathbf{x})=\lambda||\mathbf{x}||_1$, the proximity operator would have a component-wise closed-form solution which is nothing but the well known *soft-thresholding or shrinkage function* [20, Eqs. 2.34 and 2.35] given by

$$x_i{}^{(k+1)}=\operatorname{soft}(x_i{}^{(k)}-\left(\nabla h\mathbf{x}^{(k)}\right)_i,\ \lambda), \tag{3.14}$$

where $\operatorname{soft}(a,\lambda)=\operatorname{sign}(a)\max(|a|-\lambda,\ 0)$.

3.2.1 Iterative Shrinkage-Thresholding Algorithm

One of the basic forward–backward splitting algorithms is the *Iterative Shrinkage-Thresholding* (IST), where every iteration requires simple matrix-vector multiplications followed by a shrinkage or thresholding operation [21, 56]. Reconsidering the general optimization model and with exact forms of functions $h(\mathbf{x})$ and $g(\mathbf{x})$ assumed in the above example, we take the second-order approximation of $h(\mathbf{x})$ to iteratively update $\mathbf{x}$ as shown below [4, 54]

$$\begin{aligned}\mathbf{x}^{(k+1)}&=\underset{\mathbf{x}}{\operatorname{argmin}}\Big\{h\left(\mathbf{x}^{(k)}\right)+\left(\mathbf{x}-\mathbf{x}^{(k)}\right)^T\nabla h\left(\mathbf{x}^{(k)}\right)+\\&\qquad\frac{1}{2}||\mathbf{x}-\mathbf{x}^{(k)}||_2^2\nabla^2 h\left(\mathbf{x}^{(k)}\right)+\lambda g\left(\mathbf{x}\right)\Big\}\\&\approx\underset{\mathbf{x}}{\operatorname{argmin}}\left\{\frac{1}{2}||\mathbf{x}-\mathbf{v}^{(k)}||_2^2+\frac{\lambda}{\xi^{(k)}}g\left(\mathbf{x}\right)\right\}\end{aligned} \tag{3.15}$$

where $\mathbf{v}^{(k)} = \mathbf{x}^{(k)} - \frac{1}{\xi^{(k)}} \nabla h\left(\mathbf{x}^{(k)}\right)$. The Hessian matrix $\nabla^2 h\left(\mathbf{x}^{(k)}\right)$ is approximated by a diagonal matrix $I\xi^{(k)}$. Now, substituting $g(\mathbf{x}) = ||\mathbf{x}||_1$ in the above equation, results in the component-wise closed-form solution given by

$$\begin{aligned} x_i^{(k+1)} &= \underset{x}{\operatorname{argmin}} \left\{ \frac{1}{2}\left(x_i - v_i^{(k)}\right)^2 + \frac{\lambda |x_i|}{\xi^{(k)}} \right\} \\ &= \operatorname{soft}\left(v_i^{(k)}, \frac{\lambda}{\xi^{(k)}}\right) \end{aligned} \tag{3.16}$$

Algorithm 4 Iterative Shrinkage-Thresholding Algorithm

Input: $\mathbf{A}, \mathbf{y}$
Initialization: $\mathbf{x}^{(0)}, \alpha^{(0)}, \lambda_1 > \lambda_2 > \ldots. > \lambda_N$ and $k \leftarrow 0$

```
for j = 1: N do
    λ ← λ_j
    repeat
    x^(k+1) ← 𝒯†( v^(k), λ/ξ^(k) )
    ξ^(k+1) = ||A(x^(k) − x^(k−1))||_2^2 / ||x^(k) − x^(k−1)||_2^2
    k ← k + 1
    until the objective function f(x^(k)) decreases
end for
```

Output: $\mathbf{x}^* \leftarrow \mathbf{x}^{(k+1)}$
$\dagger \mathscr{T}(.)$ is the soft-thresholding operator

There are two unknown parameters in Eq. 3.16, namely, $\xi^{(k)}$ and λ. Various strategies have been proposed for selecting these parameters. Since we approximated $\nabla^2 h\left(\mathbf{x}^{(k)}\right)$ by $I\xi^{(k)}$, therefore, $\xi^{(k)}$ must follow the condition: $\xi^{(k)}\left(\mathbf{x}^{(k)} - \mathbf{x}^{(k-1)}\right) \approx \nabla h\left(\mathbf{x}^{(k)}\right) - \nabla h\left(\mathbf{x}^{(k-1)}\right)$ in least squares sense [56, p. 10], i.e.,

$$\begin{aligned} \xi^{(k+1)} &= \arg\min_{\xi} ||\xi\left(\mathbf{x}^{(k)} - \mathbf{x}^{(k-1)}\right) - \left(\nabla h\left(\mathbf{x}^{(k)}\right) - \nabla h\left(\mathbf{x}^{(k-1)}\right)\right)||_2^2 \\ &= \frac{\left(\mathbf{x}^{(k)} - \mathbf{x}^{(k-1)}\right)^T \left(\nabla h\left(\mathbf{x}^{(k)}\right) - \nabla h\left(\mathbf{x}^{(k-1)}\right)\right)}{\left(\mathbf{x}^{(k)} - \mathbf{x}^{(k-1)}\right)^T \left(\mathbf{x}^{(k)} - \mathbf{x}^{(k-1)}\right)} \end{aligned} \tag{3.17}$$

This is called the *Barzilai–Borwein equation* [2, 54]. For $h(\mathbf{x}) = \frac{1}{2}||\mathbf{A}\mathbf{x} - \mathbf{y}||_2^2$, ξ is updated as follows:

$$\xi^{(k+1)} = \frac{||A\left(\mathbf{x}^{(k)} - \mathbf{x}^{(k-1)}\right)||_2^2}{||\mathbf{x}^{(k)} - \mathbf{x}^{(k-1)}||_2^2}.$$

On the other hand, the regularization parameter λ is chosen as a decreasing sequence, i.e., first solving with a large value of λ, then continuously decreasing it for faster convergence [29, 35, 56]. The main steps of the IST algorithm are summarized in

Algorithm 4. Although its main advantage is its simplicity, but has a slower convergence rate for large-scale problems. It shows a global convergence rate of $O\left(^1/_k\right)$ [4, 20, 21], where k is the iteration counter. The IST algorithm can also be derived from the expectation maximization (EM) [28] or the majorization-minimization (MM) method [28].

3.2.2 *Two-Step Iterative Shrinkage-Thresholding Algorithm*

It has been observed that the convergence rate of the IST algorithm highly depends on the observation operator, i.e., the measurement matrix **A**. If this operator is ill-posed or ill-conditioned, then the convergence rate becomes very slow. In [9, 10], the authors proposed an algorithm known as the iterative reweighted shrinkage (IRS) which shows much faster convergence rate when **A** is strongly ill-posed. But, for mild ill-posedness of **A** and also for noisy observations, the IST converges faster than the IRS [27]. In order to exploit advantages of both the IST and the IRS, the authors in [11] proposed the Two-step IST (TwIST) algorithm that converges faster than the simple IST even when **A** is severely ill-posed. Each iteration of the TwIST is performed based on the two previous iterations. Rewriting Eq. 3.15 and then defining the general IST iteration are as follows:

$$\mathbf{x}^{(k+1)} = \underset{\mathbf{x}}{\operatorname{argmin}} \left\{ \frac{1}{2} ||\mathbf{x} - \mathbf{v}^{(k)}||_2^2 + \frac{\lambda}{\xi} g\left(\mathbf{x}\right) \right\} \tag{3.18}$$

$$\mathbf{x}^{(k+1)} = (1-\beta)\,\mathbf{x}^{(k)} + \beta\ \mathscr{T}\left(\mathbf{v}^{(k)}, \tfrac{\lambda}{\xi}\right), \tag{3.19}$$

where $\mathscr{T}(.) : \mathbb{R}^n \longrightarrow \mathbb{R}^n$ is the soft thresholding or shrinkage operator. The above equation reduces to the standard IST iteration for $\beta = 1$. Similarly, for $\beta \neq 1$, it may be treated as either under (i.e., $\beta < 1$) or over (i.e., $\beta > 1$) relaxed versions of the standard IST algorithm. The above expression shows that the next iteration $\mathbf{x}^{(k+1)}$ depends only on the previous estimate $\mathbf{x}^{(k)}$. This process may be extended further and the current estimate may be defined using two previous iterations, resulting in the ***two step*** version of the IST algorithm as follows:

$$\mathbf{x}^{(1)} = \beta\ \mathscr{T}\left(\mathbf{v}^{(0)}, \tfrac{\lambda}{\xi}\right) \tag{3.20}$$

$$\mathbf{x}^{(k+1)} = (1-\gamma)\,\mathbf{x}^{(k-1)} + (\gamma-\beta)\,\mathbf{x}^{(k)} + \beta\ \mathscr{T}\left(\mathbf{v}^{(k)}, \tfrac{\lambda}{\xi}\right), \tag{3.21}$$

where $k \geq 1$. The influence of parameters γ and β on the convergence are discussed in [11, Theorem 4]. It is fairly independent of the initial choice of γ and β. The convergence of the TwIST algorithm is two orders of magnitude faster than the standard IST. However, the computational complexity of the TwIST per iteration is more than the IST. A more detailed analysis about the TwIST can be found in [4, 11].

3.2.3 Sparse Reconstruction by Separable Approximation

Sparse reconstruction by separable approximation (SpaRSA) [54] is a general method for minimizing the sum of a smooth convex function and a non-smooth regularization function. SpaRSA is closely related to the IST algorithm. The two key differences between the IST and the SpaRSA are (1) ***selection of the step size*** $\xi^{(k)}$. In each iteration, $\xi^{(k)}$ is first obtained by the Barzilai–Borwein (BB) method in Eq. 3.17, then incremented in subsequent iterations to keep in the range $[\xi_{min}, \xi_{max}]$ until $\mathbf{x}^{(k)}$ satisfies a carefully chosen acceptance criterion and (2) ***the acceptance criterion*** - to ensure a monotonically decreasing objective function $f(\mathbf{x})$. This is because the objective function f may sometimes increase due to the Barzilai–Borwein (BB) criterion. Globally convergent BB criterion has been obtained in which the objective function is guaranteed to have a slightly smaller value than the largest value of it in the last $M + 1$ iterations [54]. Thus, an iteration is accepted only if

$$f\left(\mathbf{x}^{(k+1)}\right) \leqslant \max_{i=\max(k-M,0),\cdots,k} f\left(\mathbf{x}^{(k)}\right) - \frac{\sigma}{2}\xi^{(k)} \left\|\mathbf{x}^{(k+1)} - \mathbf{x}^{(k)}\right\|_2^2,$$

where $\sigma \in (0, 1)$ is a constant.

SpaRSA is an accelerated version of the IST due to the better selection criterion for $\xi^{(k)}$. The complete algorithm is summarized in Algorithm 5. It is also a generalized minimization method for a wide range of choices in the regularization function $g(\mathbf{x})$. For more details about the algorithm and convergence analysis, the readers may refer to [34, 54].

Algorithm 5 Sparse Reconstruction by Separable Approximation

Input: $\mathbf{y}$, $\mathbf{A}$
Initialization: $\eta > 1$, $\xi_{\min}$, $\xi_{\max} (0 < \xi_{\min} < \xi_{\max})$, $t = 0$, $k = 1$ and $\mathbf{x}^0 \leftarrow \mathbf{A}^T\mathbf{y}$

```
while not converge do
    ξ^(k) ∈ [ξ_min, ξ_max]
    while not satisfies the acceptance criterion do
        v^(k) = x^(k) − (1/ξ^(k)) ∇h(x^(k))
        x^(k+1) ← 𝒯(v^(k), λ/ξ^(k))
        ξ^(k) ← η ξ^k
    end while
    k ← k + 1
end while
```

$\mathbf{v}^{(k)} = \mathbf{x}^{(k)} - \frac{1}{\xi^{(k)}}\nabla h\left(\mathbf{x}^{(k)}\right)$; $\mathbf{x}^{(k+1)} \leftarrow \mathscr{T}\left(\mathbf{v}^{(k)}, \frac{\lambda}{\xi^{(k)}}\right)$; $\xi^{(k)} \leftarrow \eta\, \xi^k$

Output: $\mathbf{x}^* \leftarrow \mathbf{x}^{(k)}$

3.2.4 Fast Iterative Shrinkage-Thresholding Algorithm

In [4], the authors developed a new algorithm which would be computationally faster than both the IST and the TwIST whereas keeping it as simpler as possible. This algorithm has been christened as the *Fast Iterative Shrinkage-Thresholding Algorithm* (FISTA). It shows a significant improvement in the global convergence rate, both theoretically and practically. Theoretical analysis shows that the FISTA has convergence rate of the order of $O\left(1/k^2\right)$.

We recall the optimization model mentioned in Eq. 3.12,

$$\underset{\mathbf{x}}{\operatorname{argmin}}\ f(\mathbf{x}) = h(\mathbf{x}) + g(\mathbf{x}), \tag{3.22}$$

where $g : \mathbb{R}^n \rightarrow \mathbb{R}$ is a non-smooth convex function and $h : \mathbb{R}^n \rightarrow \mathbb{R}$ is a smooth convex function continuously differentiable with Lipschitz constant L, i.e.,

$$||\nabla h(\mathbf{x}) - \nabla h(\mathbf{z})|| \leq L(h)\,||\mathbf{x} - \mathbf{z}||, \quad \text{where}, \quad \mathbf{x},\, \mathbf{z} \in \mathrm{R}^{\mathrm{n}} \tag{3.23}$$

The main concept for algorithms in this category is to iteratively find quadratic approximations $Q_L(\mathbf{x}, \mathbf{z})$ of $f(\mathbf{x})$ around an appropriately chosen point $\mathbf{z}$ and then minimize $Q_L(\mathbf{x}, \mathbf{z})$ instead of $f(\mathbf{x})$. Here, $Q_L(\mathbf{x}, \mathbf{z})$ is defined as follows [4, Eq. 2.5]:

$$Q_L(\mathbf{x}, \mathbf{z}) = h(\mathbf{z}) + (\mathbf{x} - \mathbf{z})^T\, \nabla h(\mathbf{z}) + \tfrac{L}{2}||\,\mathbf{x} - \mathbf{z}||_2^2 + g(\mathbf{x}), \tag{3.24}$$

where $f(\mathbf{x}) \leq Q_L(\mathbf{x}, \mathbf{z})$ for all $\mathbf{z}$. For $g(\mathbf{x}) = \lambda||\mathbf{x}||_1$, we can write solution of the above expression as

$$\underset{\mathbf{x}}{\operatorname{argmin}}\ Q_L(\mathbf{x}, \mathbf{z}) = \mathscr{T}\left(\mathbf{v},\, \tfrac{\lambda}{L}\right), \tag{3.25}$$

where $\mathbf{v} = \mathbf{z} - \frac{1}{L}\nabla h(\mathbf{z})$. Here, $\mathbf{z}^{(k)}$ is approximated by linear combination of previous two estimates of $\mathbf{x}$, i.e., $\mathbf{x}^{(k)}$ and $\mathbf{x}^{(k-1)}$ using the following expression:

$$\mathbf{z}^{(k)} \leftarrow \mathbf{x}^{(k)} + \frac{t_{k-1} - 1}{t_k}\left(\mathbf{x}^{(k)} - \mathbf{x}^{(k-1)}\right), \tag{3.26}$$

where $\{t_k\}$ is a positive real sequence. Keeping $t_k^2 - t_k \leq t_{k-1}^2$ would achieve a convergence rate of $O\left(1/k^2\right)$ [47]. It should be noted that for large-scale problems, direct computation of Lipchitz constant L may become expensive which may be mitigated by applying the backtracking line search technique.

As discussed above, differences between the IST and the FISTA are first, FISTA estimates the shrinkage operation $\mathscr{T}(\cdot)$ based on $\mathbf{z}^{(k)}$ instead of $\mathbf{x}^{(k)}$. Second, in order to achieve acceleration in the convergence rate of FISTA, $\mathbf{z}^{(k)}$ is obtained by linear combination of previous two points, $\left\{\mathbf{x}^{(k)}, \mathbf{x}^{(k-1)}\right\}$. The FISTA is summarized in Algorithm 6. Out of all the above operator splitting techniques, FISTA has the fastest convergence rate. Computational complexity per iteration of the FISTA is similar to the IST algorithm. In the FISTA, the current iteration is obtained from linear combination of two previous iterations followed by a shrinkage operation. Similarly, each iteration of the IST algorithm also involves a shrinkage operation. So, the main computational cost of these two algorithms is due to the shrinkage operation only neglecting the additional steps for updation of the iteration and the step size in the FISTA. However, the TwIST contains an extra shrinkage operation outside the main loop besides the one inside it, so the per-iteration cost of the TwIST is slightly more than that of the IST or the FISTA. Similar analysis also reveals that the SparSA bears computational cost as that of the IST.

In parallel to the development of FISTA, a very similar algorithm was introduced, namely, the NESTA (Nesterov's work) by Becker et al. [5]. It has also the same global convergence rate $O\left(1/k^2\right)$. It is mainly based on the works of [47].

3.2.5 *Total Variation ℓ_1 Compressed MR Imaging*

Total variation ℓ_1 compressed MR imaging (TVCMRI) is an efficient algorithm for the CS-based MR image reconstruction using total variation (TV) and wavelets [43]. The key differences between the above-discussed algorithms and the TVCMRI can be summarized in the following points:

1. MR images contain different regions which are piecewise smooth. In order to include this behavior of MR images, the TVCMRI algorithm considers an additional regularization term based on total variation within the model given in Eq. 3.12.
2. Unlike the IST and the FISTA, in TVCMRI, the joint optimization model with the ℓ_1 and the TV regularizations, is solved by two operator splitting stages iteratively.

Algorithm 6 Fast Iterative Shrinkage-Thresholding Algorithm

Input: $\mathbf{A},\ \mathbf{y},\ \lambda_1$
Initialization: $L \leftarrow L\left(h\right),\ \mathbf{x}^{(0)} \leftarrow \mathbf{A}^T\mathbf{y},\ \mathbf{x}^{(1)} \leftarrow \mathbf{A}^T\mathbf{y},\ t_0 \leftarrow 1, t_1 \leftarrow 1,\ k \leftarrow 1,\ \beta \in (0,1)$ and $\hat{\lambda} > 0$

while not converge **do**
  $\mathbf{z}^{(k)} \leftarrow \mathbf{x}^{(k)} + \frac{t_{k-1}-1}{t_k}\left(\mathbf{x}^{(k)} - \mathbf{x}^{(k-1)}\right)$
  $\mathbf{v}^{(k)} \leftarrow \mathbf{z}^{(k)} - \frac{1}{L}\nabla f\left(\mathbf{z}^{(k)}\right)$
  $\mathbf{x}^{(k+1)} \leftarrow \mathscr{T}\left(\mathbf{v}^{(k)},\ \frac{\lambda_k}{L}\right)$
  $t_{k+1} \leftarrow \frac{1+\sqrt{4t_k^2+1}}{2}$
  $\lambda_{k+1} \leftarrow \max\left(\beta\lambda_k,\ \hat{\lambda}\right)$
  $k \leftarrow k+1$
end while
Output: $\mathbf{x}^* \leftarrow \mathbf{x}^{(k)}$

Better acceleration and convergence can be achieved as the intermediate results are continuously updated by the dual operator splitting stages.

To reconstruct MR image from the measured data, consider the problem P_2 (see Chap. 2).

$$\hat{\mathbf{x}} = \arg\min_{\mathbf{x}} f(\mathbf{x}) = h(\mathbf{x}) + \lambda_1\ g_1(\mathbf{x}) + \lambda_2\ g_2^+(\mathbf{x}), \tag{3.27}$$

where $h(\mathbf{x}) = \|\mathbf{A}\mathbf{x} - \mathbf{y}\|_2^2$, $g_1(\mathbf{x}) = \|\Psi\mathbf{x}\|_1$ and $g_2^+(\mathbf{x}) = ||\mathbf{x}||_{TV}$. The TV-$\ell_1$-$\ell_2$ model for MR image reconstruction was also applied by Lustig et al. [42]. As both the regularization terms TV and ℓ_1 -norms are non-smooth, this model is more difficult to solve than either the ℓ_1-ℓ_2 or the TV-ℓ_2 model. Since all the terms in (3.27) are convex and $\lambda_1,\ \lambda_2 > 0$, the objective function $f(\mathbf{x})$ is also convex. We now define $\mathbf{D} \in \mathbb{R}^{2n\times n} = \left(\mathbf{D}^{(1)}; \mathbf{D}^{(2)}\right)$, where $\left\{\mathbf{D}^{(j)}\right\}_{j=1,2} \in R^{n\times n}$ are the two first-order discrete finite difference operators in horizontal and vertical directions. Using the equivalent notation for the TV norm regularization function $g_2^+(\mathbf{x}) = ||\mathbf{x}||_{TV} = ||\mathbf{D}(\mathbf{x})||_2 = g_2(\mathbf{D}(\mathbf{x}))$, where $g_2(.) = ||.||_2$, we can write the first-order optimality condition of the above problem as

$$\mathbf{0} \in \partial f\left(\mathbf{x}^*\right) = \nabla h\left(\mathbf{x}^*\right) + \lambda_1 \partial g_1\left(\mathbf{x}^*\right) + \lambda_2 \sum_{i=1}^{n} \partial g_2\left((\mathbf{D}\mathbf{x}^*)_i\right), \tag{3.28}$$

where $\partial f(\mathbf{x}^*)$ is the set of sub-gradients of f at $\mathbf{x}^*$. Now, we can apply the general property for any convex function f and its convex conjugate, i.e.,

$$\mathbf{y} \in \partial f(\mathbf{x}) \quad \Leftrightarrow \quad \mathbf{x} \in \partial f^*(\mathbf{y}). \tag{3.29}$$

Equivalently, $\mathbf{x}^*$ is optimal if and only if there is an auxiliary variable $\mathbf{v}^* = \left\{v_i^*\right\}$ where $v_i \in \mathbb{R}^2$, i.e.,

$$\mathbf{0} \in \nabla_x h\left(\mathbf{x}^*\right) + \lambda_1 \partial g_1\left(\mathbf{x}^*\right) + \lambda_2 \sum_{i=1}^{n} \mathbf{D}_i^* v_i^* \tag{3.30}$$

$$\mathbf{D}_i \mathbf{x}^* \in \partial g_2^*\left(v_i^*\right), \tag{3.31}$$

where $\mathbf{D}_i \in \mathbb{R}^{2\times n}$ finds discrete finite differences in horizontal and vertical directions at the ith pixel of the image and $\mathbf{D}_i^*$ denotes the transpose of $\mathbf{D}_i$. To apply the operator splitting technique, slight rearrangements of the above equations are to be carried out as follows:

$$\mathbf{0} \in \tau_1 \lambda_1 \partial g_1\left(\mathbf{x}^*\right) + \mathbf{x}^* - \mathbf{s} \tag{3.32}$$

$$\mathbf{s} = \mathbf{x}^* - \tau_1 \left(\nabla_x\ h\left(\mathbf{x}^*\right) + \lambda_2 \sum_{i=1}^{n} \mathbf{D}_i^* v_i^*\right) \tag{3.33}$$

$$\mathbf{0} \in \tau_2 \lambda_2 \partial g_2^*\left(v_i^*\right) + v_i^* - t_i \tag{3.34}$$

$$t_i = v_i^* + \tau_2 \mathbf{D}_i \mathbf{x}_i^*. \tag{3.35}$$

Now, for given $\mathbf{x}^*$ and v_i^*, it is easy and straightforward to compute $\mathbf{s}$ and t_i. On the other side, for given $\mathbf{s}$ and t_i, one can uniquely determine $\mathbf{x}^*$ and v_i^* using backward step of the operator splitting technique as given below [43, see Eq. 15 and 17].

$$x_i^*(s) = \text{soft}\,(s_i,\ \tau_1\lambda_1) \tag{3.36}$$

and

$$v_i^* = \min\left\{\frac{1}{\tau_1\lambda_2},\ \| \ t_i \ \|_2\right\} \frac{t_i}{\| \ t_i \ \|_2}, \tag{3.37}$$

where every operation is performed component-wise and i represents the location of the component. TVCMRI is a four-step iteration-based algorithm, namely, two forward steps (Eqs. 3.33 and 3.35) and two backward steps (Eqs. 3.36 and 3.37). Solutions of these equations are guaranteed to converge with finite number of iterations when the step size τ_1 and τ_2 are sufficiently small [20]. The algorithm is quite faster for recovering the MR image compared to the iterative shrinkage algorithms.

3.3 Variable Splitting Method

The variable splitting technique takes a divide and conquer approach to solve a complex optimization problem. This is done by replacing the problem of estimating a variable of interest with a sequence of subproblems through the introduction of new variables [24]. Then solutions of these additional variables are used to estimate the original variable of interest. This approach primarily reduces the computational complexity because the subproblems are much simpler to solve compared to the original problem. Moreover, the subproblems are generally solved by simple existing optimization techniques [1, 12]. For example, consider an unconstrained optimization problem

Algorithm 7 Total variation ℓ_1 compressed MR imaging (TVCMRI)

Input: $\mathbf{y},\ \mathbf{A},\ \Psi,\ \mathbf{D}$

Initialization: $(\lambda_1,\ \lambda_2,\ \tau_1,\ \tau_2) > 0,\ k \leftarrow 0,\ \mathbf{v}^{(0)} \leftarrow \mathbf{0}$, and $\mathbf{x}^{(0)} \leftarrow \mathbf{A}^T\mathbf{y}$

1: **while** not converge **do**

2: $\quad \mathbf{s}^{(k+1)} \leftarrow \mathbf{x}^{(k)} - \tau_1\left(A^T\left(A\mathbf{x}^{(k)} - \mathbf{y}\right) + \lambda_2\mathbf{D}^T\mathbf{v}^{(k)}\right)$

3: $\quad \mathbf{t}^{(k+1)} \leftarrow \mathbf{v}^{(k)} + \tau_2\mathbf{D}\mathbf{x}^{(k)}$

4: $\quad \mathbf{x}^{(k+1)} \leftarrow \mathscr{T}\left(\mathbf{s}^{(k+1)},\ \tau_1\lambda_1\right)$

5: $\quad \mathbf{v}^{(k+1)} \leftarrow \min\left\{\frac{1}{\tau_1\lambda_2},\ \| \ \mathbf{t}^{(k+1)} \ \|_2\right\} \frac{\mathbf{t}^{(k+1)}}{\|\mathbf{t}^{(k+1)}\|_2}$

6: $\quad k \leftarrow k+1$

7: **end while**

Output: $\mathbf{x}^* \leftarrow \mathbf{x}^{(k)}$

$$\underset{\mathbf{x}}{\text{minimize}} \quad f_1(\mathbf{x}) + f_2(g(\mathbf{x})), \tag{3.38}$$

where $g : \mathbb{R}^n \rightarrow \mathbb{R}^n$. Clearly, $\mathbf{x}$ is coupled with several other variables and solving the above problem as a whole to estimate $\mathbf{x}$ is very complex. However, a simplification can be achieved by replacing (3.38) with a sequence of subproblems through introduction of a new variable $\mathbf{v}$ such that $\mathbf{v} = g(\mathbf{x})$, i.e., by writing

$$\begin{aligned} &\underset{\mathbf{x},\mathbf{v}}{\text{minimize}} && f_1(\mathbf{x}) + f_2(\mathbf{v}) \\ &\text{subject to} && g(\mathbf{x}) = \mathbf{v} \end{aligned} \tag{3.39}$$

By augmented Lagrangian formulation for variable splitting (to be detailed later), we can easily solve this problem using the following two steps:

$$\left(\mathbf{x}^{k+1}, \mathbf{v}^{k+1}\right) \in \underset{\mathbf{x},\mathbf{v}}{\text{argmin}} \quad f_1(\mathbf{x}) + f_2(\mathbf{v}) + \tfrac{\mu}{2} \| g(\mathbf{x}) - \mathbf{v} - \mathbf{d}^k \|_2^2 \tag{3.40}$$

$$\mathbf{d}^{k+1} = \mathbf{d}^k - \left(g\left(\mathbf{x}^{k+1}\right) - \mathbf{v}^{k+1}\right). \tag{3.41}$$

The minimization problem in Eq. (3.40) can be alternately minimized with respect to $\mathbf{x}$ and $\mathbf{v}$ followed by updating $\mathbf{d}$ until convergence using Eq. (3.41).

Next, we bring out a detailed discussion about some of the well-known variable splitting algorithms that are used very frequently in CS-MRI.

3.3.1 Augmented Lagrange Multiplier Method

Consider an equality constrained optimization problem:

$$\begin{aligned} &\underset{\mathbf{x}}{\text{minimize}} && f(\mathbf{x}) \\ &\text{subject to} && h(\mathbf{x}) = 0, \end{aligned} \tag{3.42}$$

where $f : \mathbb{R}^n \rightarrow R$ and $h : \mathbb{R}^n \rightarrow \mathbb{R}^m$ are given objective function and constraints, respectively. For example, assume that $f(\mathbf{x}) = ||\mathbf{x}||_1$ and $h(\mathbf{x}) = \mathbf{y} - \mathbf{A}\mathbf{x}$. Lagrangian method converts the above-constrained problem into an unconstrained problem, i.e.,

$$L(\mathbf{x}, \boldsymbol{\lambda}) = f(\mathbf{x}) + \boldsymbol{\lambda}^T h(\mathbf{x}) \tag{3.43}$$

where $\boldsymbol{\lambda} \in \mathbb{R}^m$ is the Lagrange multiplier vector. This problem may be solved by the gradient descent method where both $\mathbf{x}$ and λ are updated iteratively until convergence with the assumption of local convexity condition [7, Ch. 1, Eq. 2]. Alternatively, an ascent method may also be adopted to maximize the dual of Eq. 3.43, given by

$$d(\boldsymbol{\lambda}) = \inf_{\mathbf{x}} \left\{ f(\mathbf{x}) + \boldsymbol{\lambda}^T h(\mathbf{x}) \right\} = \inf_{\mathbf{x}} \left\{ L(\mathbf{x}, \boldsymbol{\lambda}) \right\}. \tag{3.44}$$

Maximization of the above dual function gives the following rule for updating $\boldsymbol{\lambda}$. That is

$$\boldsymbol{\lambda}^{(k+1)} = \boldsymbol{\lambda}^{(k)} + \gamma\, h\left(\mathbf{x}^{(k)}\right), \tag{3.45}$$

where $\mathbf{x}^{(k)}$ is solution of the primal problem, i.e., the minimization of $L\left(\mathbf{x}^{(k)}, \boldsymbol{\lambda}^{(k)}\right)$ and γ is the fixed step size scalar parameter. The above method is called the primal-dual method. The method assumes that $L\left(\mathbf{x}^{*}, \boldsymbol{\lambda}^{*}\right)$ satisfies the local convexity condition, i.e., $\nabla^2 L(\mathbf{x}^{*}, \boldsymbol{\lambda}^{*}) > 0$ which may not be always possible [7, Ch. 1, Eq. 2]. Moreover, it is also affected by the slow convergence and insufficient *a priori* information about the step size γ.

A different approach to convert the problem in Eq. 3.42 to an unconstrained optimization problem is the penalty function method. In this method, the above constrained optimization problem may be formulated as

$$\begin{aligned} &\underset{\mathbf{x}}{\text{minimize}} \quad f\left(\mathbf{x}\right) + \tfrac{1}{2} c_k \|h\left(\mathbf{x}\right)\|^2 \\ &\text{subject to} \;\; \mathbf{x} \in \mathbb{R}^n \end{aligned} \tag{3.46}$$

where c_k is a positive and continually increasing sequence for all k and $c_k \to \infty$. Minimization of the above problem requires the solution of

$$\inf_{\mathbf{x}\in\mathbb{R}^n} \lim_{c_k\to\infty} \left\{ f\left(\mathbf{x}\right) + \tfrac{1}{2} c_k \|h\left(\mathbf{x}\right)\|^2 \right\}. \tag{3.47}$$

The penalty method is widely accepted in practice due to its simplicity, the ability to deal nonlinear constrained problem and availability as a powerful unconstrained minimization approach [7, Chap. 1]. But, it has also some limitations like (1) slow convergence and (2) ill-conditioning for large values of c_k.

Hestenes [36] and Powell [51] proposed the *method of multipliers also known as the augmented Lagrangian method* where the idea of penalty method is merged with those of the primal-dual and basic Lagrangian approaches. In this approach, a quadratic penalty term is added to the Lagrangian function in Eq. 3.43. Thus, the new format for the objective function is

$$L_c\left(\mathbf{x}, \boldsymbol{\lambda}\right) = f\left(\mathbf{x}\right) + \boldsymbol{\lambda}^T h\left(\mathbf{x}\right) + \tfrac{1}{2} c \|h\left(\mathbf{x}\right)\|^2, \tag{3.48}$$

and associated sequence of minimizations and updations to be taken are

$$1: \begin{aligned} &\text{minimize} \quad L_{c_k}\left(\mathbf{x}, \boldsymbol{\lambda}^{(k)}\right) \\ &\text{subject to} \;\; \mathbf{x} \in \mathbb{R}^n \end{aligned}, \tag{3.49}$$

and

$$2: \boldsymbol{\lambda}^{(k+1)} = \boldsymbol{\lambda}^{(k)} + c_k h\left(\mathbf{x}^{(k)}\right). \tag{3.50}$$

The above two-step formulation is used for solving any convex optimization problem using the *augmented Lagrangian multiplier* (ALM) method.

ALM Formulation for Variable Splitting

Consider an unconstrained optimization problem where the main objective function can be split into two different functions and out of these two functions, one could be written as the composition of two functions, i.e.,

$$\underset{\mathbf{x}}{\text{minimize}} \quad f_1(\mathbf{x}) + f_2(g(\mathbf{x})). \tag{3.51}$$

Now, following the assumptions reported in [1, Sec. II], and rewriting the above minimization problem by introducing a new variable $\mathbf{v}$ as

$$\begin{aligned} \underset{\mathbf{x}\in\mathbb{R}^n, \mathbf{v}\in\mathbb{R}^n}{\text{minimize}} \quad & f_1(\mathbf{x}) + g_2(\mathbf{v}) \\ \text{subject to} \quad & g(\mathbf{x}) = \mathbf{v} \end{aligned} \tag{3.52}$$

Using the steps of ALM (Eqs. 3.49 and 3.50), we can minimize the above problem by the following steps:

$$\begin{aligned} & \underset{\mathbf{x}\in\mathbb{R}^n, \mathbf{v}\in\mathbb{R}^n}{\text{minimize}} \quad L_{c_k}\left(\mathbf{x}, \mathbf{v}, \boldsymbol{\lambda}^{(k)}\right) \\ & \boldsymbol{\lambda}^{(k+1)} = \boldsymbol{\lambda}^{(k)} - c_k\left(g(\mathbf{x}) - \mathbf{v}\right) \end{aligned} \tag{3.53}$$

where $L_{c_k}\left(\mathbf{x}, \mathbf{v}, \boldsymbol{\lambda}^{(k)}\right) = f_1(\mathbf{x}) + f_2(\mathbf{v}) - \boldsymbol{\lambda}^{(k)^T}\left(g(\mathbf{x}) - \mathbf{v}\right) + \frac{c_k}{2}||g(\mathbf{x}) - \mathbf{v}||_2^2$. Simplifying this expression, we get

$$\begin{aligned} L_{c_k}\left(\mathbf{x}, \mathbf{v}, \boldsymbol{\lambda}^{(k)}\right) &= f_1(\mathbf{x}) + f_2(\mathbf{v}) + \frac{c_k}{2}\left(||g(\mathbf{x}) - \mathbf{v}||_2^2 - \frac{2}{c_k}\boldsymbol{\lambda}^{(k)^T}\left(g(\mathbf{x}) - \mathbf{v}\right) + \right. \\ & \qquad \left. \left(\frac{\boldsymbol{\lambda}^{(k)}}{c_k}\right)^2 - \left(\frac{\boldsymbol{\lambda}^{(k)}}{c_k}\right)^2\right) \\ &= f_1(\mathbf{x}) + f_2(\mathbf{v}) + \frac{c_k}{2}\left(||g(\mathbf{x}) - \mathbf{v} - \frac{\boldsymbol{\lambda}^{(k)}}{c_k}||_2^2 - \left(\frac{\boldsymbol{\lambda}^{(k)}}{c_k}\right)^2\right). \end{aligned} \tag{3.54}$$

Now, we may also neglect the terms independent of $\mathbf{x}$ and $\mathbf{v}$ in the above expression while performing joint minimization with respect to $\mathbf{x}$ and $\mathbf{v}$. Thus,

$$L_{c_k}\left(\mathbf{x}, \mathbf{v}, \mathbf{d}^{(k)}\right) = f_1(\mathbf{x}) + f_2(\mathbf{v}) + \frac{c_k}{2}||g(\mathbf{x}) - \mathbf{v} - \mathbf{d}^{(k)}||_2^2, \tag{3.55}$$

where $\mathbf{d}^{(k)} = \frac{\boldsymbol{\lambda}^{(k)}}{c_k}$. Updates of the sequence $\left\{\mathbf{d}^{(k)}\right\}$ are given by:

$$\begin{aligned} \mathbf{d}^{(k+1)} &= \frac{\boldsymbol{\lambda}^{(k+1)}}{c_k} \\ &= \frac{\boldsymbol{\lambda}^{(k)} - c_k\left(g(\mathbf{x}^{(k+1)}) - \mathbf{v}^{(k+1)}\right)}{c_k} \end{aligned}$$

$$
\begin{aligned}
&= \frac{\lambda^{(k)}}{c_k} - \left(g(\mathbf{x}^{(k+1)}) - \mathbf{v}^{(k+1)}\right) \\
&= \mathbf{d}^{(k)} - \left(g(\mathbf{x}^{(k+1)}) - \mathbf{v}^{(k+1)}\right). \qquad (3.56)
\end{aligned}
$$

Finally, using the steps of the nonlinear block–Gauss–Seidel (NLBGS) technique [8, p. 267], joint minimization of the problem in (3.55) is to be carried out with respect to $\mathbf{x}$ and $\mathbf{v}$ for a given value of $\mathbf{d}^{(k)}$ as follows:

$$
\begin{aligned}
&\text{Step a: Determine } \mathbf{x}^{(k+1)} = \underset{\mathbf{z}_1 \in \mathbf{x}}{\operatorname{argmin}} \; L_{c_k}\left(\mathbf{z}_1, \mathbf{v}^{(k)}, d^{(k)}\right) \\
&\text{Step b: Determine } \mathbf{v}^{(k+1)} = \underset{\mathbf{z}_2 \in \mathbf{v}}{\operatorname{argmin}} \; L_{c_k}\left(\mathbf{x}^{(k+1)}, \mathbf{z}_2, d^{(k)}\right) \text{ such that} \\
&\quad L_{c_k}\left(\mathbf{x}^{(k+1)}, \mathbf{v}^{(k+1)}, d^{(k)}\right) \leqslant L_{c_k}\left(\mathbf{x}^{(k+1)}, \mathbf{v}^{(k)}, d^{(k)}\right) \text{ and} \\
&\quad \nabla_{\mathbf{v}} L_{c_k}\left(\mathbf{x}^{(k+1)}, \mathbf{v}^{(k+1)}, d^{(k)}\right)^T \left(\mathbf{z}_2 - \mathbf{v}^{(k+1)}\right) \geqslant 0 \text{ for all } \mathbf{z}_2 \in \mathbf{v}
\end{aligned} \qquad (3.57)
$$

For convergence, one needs to solve the above steps with good accuracy before updating $\mathbf{d}^{(k)}$ which makes each iteration of the classical ALM technique quite expensive. We now summarize the steps of ALM with variable splitting in Algorithm 8.

Algorithm 8 Augmented Lagrange Multiplier(ALM) Methods

Initialization: $k \leftarrow 0,\ \mathbf{x} \leftarrow \mathbf{0},\ \mathbf{v} \leftarrow \mathbf{0},\ \mathbf{d}^0$ and $c_k > 0$

1: **while** not converged **do**

2: $\quad \left(\mathbf{x}^{(k+1)}, \mathbf{v}^{(k+1)}\right) \leftarrow \underset{\mathbf{x},\mathbf{v}}{\operatorname{argmin}}\ f_1(\mathbf{x}) + f_2(\mathbf{v}) + \frac{c_k}{2} ||g(\mathbf{x}) - \mathbf{v} - \mathbf{d}^{(k)}||_2^2$

3: $\quad \mathbf{d}^{(k+1)} \leftarrow \mathbf{d}^{(k)} - \left(g(\mathbf{x}^{(k+1)}) - \mathbf{v}^{(k+)}\right)$

4: $\quad k \leftarrow k + 1$

5: **end while**

Output: $\mathbf{x}^* \leftarrow \mathbf{x}^{(k)}$

3.3.2 Alternating Direction Method of Multipliers

The alternating direction method of multipliers (ADMM) is quite similar to the ALM. The original idea of the ADMM comes from the works of Gabay and Mercier [30, 32]. Glowinski and Tallec [31] also interpreted the ADMM as the Douglas–Rachford splitting method. Further, equivalence of the proximal method and the ADMM is discussed in the works of Eckstein and Bertsekas [26].

Let us start with the framework reported in [57, Sec. 2] and [1, Sec. II(C)] based on the original works of Gabay and Mercier. In order to avoid the computational complexity that exists in the joint minimization of the problem in (3.57) using the ALM, we may solve the same problem by dividing it into two comparatively simpler subproblems which are to be minimized separately. This idea leads to the development of the *alternating direction method of multipliers* (ADMM). In particular, ADMM minimizes $L_{c_k}\left(\mathbf{x}, \mathbf{v}, \mathbf{d}^{(k)}\right)$ with respect to $\mathbf{x}$ and $\mathbf{v}$ separately by one NLBGS iteration. After every alternating minimization of $L_{c_k}(\mathbf{x}, \mathbf{v}, \mathbf{d})$ with respect to $\mathbf{x}$ and $\mathbf{v}$, the sequence $\mathbf{d}^{(k)}$ is updated immediately. Thus, for the ADMM following three steps are executed sequentially:

$$\left.\begin{aligned} \mathbf{x}^{(k+1)} &= \underset{\mathbf{x}}{\operatorname{argmin}}\ L_{c_k}\left(\mathbf{x}, \mathbf{v}^{(k)}, \mathbf{d}^{(k)}\right) \\ \mathbf{v}^{(k+1)} &= \underset{\mathbf{v}}{\operatorname{argmin}}\ L_{c_k}\left(\mathbf{x}^{(k+1)}, \mathbf{v}, \mathbf{d}^{(k)}\right) \\ \mathbf{d}^{(k+1)} &= \mathbf{d}^{(k)} - \left(g\left(\mathbf{x}^{(k+1)}\right) - \mathbf{v}^{(k+1)}\right). \end{aligned}\right\} \tag{3.58}$$

From the above, we may conclude that each step of the ADMM is relatively much cheaper than the ALM.

Convergence analysis of the ADMM algorithm was carried out by Afonso et al. in [1, Theorem 1] and Eckstein and Bertsekas in [26, Theorem 8]. According to them, $\mathbf{x}^{(k)}$ and $\mathbf{v}^{(k)}$ should satisfy the following conditions for the convergence:

$$\eta^{(k)} \geq \|\mathbf{x}^{(k+1)} - \underset{\mathbf{x}}{\operatorname{argmin}}\{f_1(\mathbf{x}) + \tfrac{c_k}{2}\|g(\mathbf{x}) - \mathbf{v} - \mathbf{d}^{(k)}\|_2^2\}\| \tag{3.59}$$

$$\nu^{(k)} \geq \|\mathbf{v}^{(k+1)} - \underset{\mathbf{x}}{\operatorname{argmin}}\{f_2(\mathbf{v}) + \tfrac{c_k}{2}\|g(\mathbf{x}) - \mathbf{v} - \mathbf{d}^{(k)}\|_2^2\}\| \tag{3.60}$$

$$\mathbf{d}^{(k+1)} = \mathbf{d}^{(k)} - \left(g\left(\mathbf{x}^{(k+1)}\right) - \mathbf{v}^{(k+1)}\right), \tag{3.61}$$

where $\{\eta_k \geq 0,\ k = 0, 1, \ldots\}$ and $\{\nu_k \geq 0,\ k = 0, 1, \ldots\}$ be two sequences such that $\sum_{k=0}^{\infty} \eta_k < \infty$ and $\sum_{k=0}^{\infty} \nu_k < \infty$.

Comparisons of the above three conditions to the sequence of minimizations in (3.58), clearly explain that convergence of the ADMM is guaranteed even if the problems in (3.58) could not be solved exactly provided that the corresponding errors are absolutely summable. If these errors are zero, then we would end up with solutions which are exactly same to that obtained from minimizations in problem (3.58). Thus, from the convergence point of view, the ALM algorithm requires somewhat stringent conditions than that of the ADMM. Some of the very popular algorithms in CS-MRI under the ADMM category are discussed below:

3.3.2.1 Split Augmented Lagrangian Shrinkage Algorithm

Split augmented Lagrangian shrinkage algorithm (SALSA) [1] is based on the idea of variable splitting technique. After splitting, each subproblem is minimized by the ADMM technique.

Let us now turn to our unconstrained optimization model in (3.38), i.e.,

$$\underset{\mathbf{x} \in \mathbb{R}^n}{\operatorname{minimize}}\ \ f_1(\mathbf{x}) + f_2(g(\mathbf{x})), \tag{3.62}$$

where $f_2(\mathbf{x})$ may be either sparsity based or the TV based regularization prior. For example, if $f_1(\mathbf{x}) = \frac{1}{2}||\mathbf{Ax} - \mathbf{y}||_2^2$, and $f_2(\mathbf{x})$ represents the TV norm regularizer, i.e., $g(\mathbf{x}) = \mathbf{Dx}$. Now, applying the variable splitting technique on the above problem, we may rewrite the problem as follows:

$$\begin{aligned} &\underset{\mathbf{x},\mathbf{v}\,\in R^n}{\text{minimize}} \quad \tfrac{1}{2}||\mathbf{Ax} - \mathbf{y}||_2^2 + f_2(\mathbf{v}) \\ &\text{subject to} \quad g(\mathbf{x}) = \mathbf{v}. \end{aligned} \tag{3.63}$$

Thus, the above problem is the same as Eq. 3.52. Therefore, we may replace the above problem with the iterative minimization of its equivalent augmented lagrangian function $L_{c_k}\left(\mathbf{x}, \mathbf{v}, \mathbf{d}^{(k)}\right)$ along with the update of $\left\{\mathbf{d}^{(k)}\right\}$. So, we write

$$L_{c_k}\left(\mathbf{x}, \mathbf{v}, \mathbf{d}^{(k)}\right) = \tfrac{1}{2}\left\|\mathbf{Ax} - \mathbf{y}\right\|_2^2 + \tau g(\mathbf{v}) + \tfrac{c_k}{2}\left\|\mathbf{x} - \mathbf{v} - \mathbf{d}^{(k)}\right\|_2^2 \tag{3.64}$$

$$\mathbf{d}^{(k+1)} = \mathbf{d}^{(k)} - (\mathbf{x}^{(k+1)} - \mathbf{v}^{(k+1)}), \tag{3.65}$$

where $\mathbf{d}^{(k)} = \frac{\lambda^{(k)}}{c_k}$. Minimization of $L_{c_k}\left(\mathbf{x}, \mathbf{v}, \mathbf{d}^{(k)}\right)$ using the ADMM algorithm would require solution of the following steps in a sequential manner.

$$\mathbf{x}^{(k+1)} = \underset{\mathbf{x}}{\text{argmin}}\ ||\mathbf{Ax} - \mathbf{y}||_2^2 + \mu\,||\mathbf{x} - \mathbf{v}^{(k)} - \mathbf{d}^{(k)}||_2^2 \tag{3.66}$$

$$\mathbf{v}^{(k+1)} = \underset{\mathbf{v}}{\text{argmin}}\ \tau\ g(\mathbf{v}) + \frac{\mu}{2}\,||\mathbf{x}^{(k+1)} - \mathbf{v} - \mathbf{d}^{(k)}||_2^2 \tag{3.67}$$

$$\mathbf{d}^{(k+1)} = \mathbf{d}^{(k)} - \left(\mathbf{x}^{(k+1)} - \mathbf{v}^{(k+1)}\right). \tag{3.68}$$

The above equations define the three main steps of the SALSA. Similar to the ADMM, convergence is guaranteed if $\mathbf{x}^{(k)}$ and $\mathbf{v}^{(k)}$ satisfies the conditions given by Eqs. (3.59)–(3.61). Now, an inspection of Eq. 3.66 confirms the minimization of a strictly convex quadratic function. So, differentiating it with respect to $\mathbf{x}$ gives an exact solution of $\mathbf{x}$, i.e.,

$$\mathbf{x}^{(k+1)} = \left(\mathbf{A}^T\mathbf{A} + \mu\mathbf{I}\right)^{-1}\left(\mathbf{A}^T\mathbf{y} + \mu\,\mathbf{x}'^{(k)}\right), \tag{3.69}$$

where $\mathbf{x}'^{(k)} = \mathbf{v}^{(k)} + \mathbf{d}^{(k)}$.

We observe that the term $\left(\mathbf{A}^T\mathbf{A} + \mu\mathbf{I}\right)$ is the regularized version of the Hessian matrix $\mathbf{A}^T\mathbf{A}$. For particular choices of $\mathbf{A}$, inverse of the matrix $\left(\mathbf{A}^T\mathbf{A} + \mu\mathbf{I}\right)$ can be done accurately [1, see Sec. III, Part-B]. Also, the computational cost of the matrix inversion step $\left(\mathbf{A}^T\mathbf{A} + \mu\mathbf{I}\right)^{-1}$ is of the order of $O(n \log n)$ for above selections of $\mathbf{A}$. Thus, SALSA makes use of the second-order information of the data fidelity term in a very efficient way. This is very different from the IST algorithms where information of only first-order differentiation of the data fidelity term is used, i.e., approximating the Hessian matrix by $(\mathbf{I}\,L(h))$ [4]. Therefore, we conclude that the SALSA could be a very good choice for particular applications where computation of $\left(\mathbf{A}^T\mathbf{A}\right)$ can be done efficiently such that its inversion is feasible. The main steps of the SALSA algorithm are now summarized in Algorithm 9. For more details about the algorithm, the interested reader may refer to [1].

Algorithm 9 Split Augmented Lagrange Shrinkage Algorithm

Input: $\mathbf{y}$, $\mathbf{A}$, $\mathbf{v}^{(0)}$, and $\mathbf{d}^{(0)}$
Initialization: $k \leftarrow 0$, $\mathbf{x}^{(0)} \leftarrow \mathbf{A}^T\mathbf{y}$, τ, and $\mu > 0$
1: **while** not converge **do**
2: $\mathbf{x}^{(k+1)} \leftarrow \underset{\mathbf{x}}{\text{argmin}} \ ||\mathbf{A}\mathbf{x} - \mathbf{y}||_2^2 + \mu\, ||\mathbf{x} - \mathbf{v}^{(k)} - \mathbf{d}^{(k)}||_2^2$
3: $\mathbf{v}^{(k+1)} \leftarrow \underset{\mathbf{v}}{\text{argmin}} \ \tau\ g(\mathbf{v}) + \frac{\mu}{2}\, ||\mathbf{x}^{(k+1)} - \mathbf{v} - \mathbf{d}^{(k)}||_2^2$
4: $\mathbf{d}^{(k+1)} \leftarrow \mathbf{d}^{(k)} - \left(\mathbf{x}^{(k+1)} - \mathbf{v}^{(k+1)}\right)$
5: $k \leftarrow k + 1$
6: **end while**
Output: $\mathbf{x}^* \leftarrow \mathbf{x}^{(k)}$

3.3.2.2 Reconstruction from Partial Fourier Data (RecPF)

Reconstruction from partial Fourier data (RecPF) [58] algorithm is based on the ADMM, developed especially for signal reconstruction from partial Fourier measurements. This is the first ADMM-based variable splitting algorithm for solving an unconstrained optimization problem with composite regularization terms, i.e., having both TV and ℓ_1 regularizations. It has faster convergence rate compared to the previously discussed TV-ℓ_1-ℓ_2 algorithm, i.e., the TVCMRI in the operator splitting category as reported in [58, Section III(C)].

We proceed by defining $\mathbf{F}_u = \boldsymbol{\Phi}\mathbf{F}$, where $\mathbf{F}_u$ is the $m \times n$ Fourier undersampling operator, $\mathbf{F}$ is the $n \times n$ Fourier transform matrix and $\boldsymbol{\Phi}$ is an $m \times n$ matrix formed by m rows of an $n \times n$ identity matrix. Now, we recall P2 problem [see Chapter 2], i.e.,

$$\hat{\mathbf{x}} = \underset{\mathbf{x}}{\text{argmin}} \sum_i ||\mathbf{D}_i\mathbf{x}||_2 + \tau||\,\Psi\mathbf{x}||_1 + \frac{\mu}{2}||\mathbf{F}_u\mathbf{x} - \mathbf{y}||_2^2, \tag{3.70}$$

where $\mathbf{D}_i \in \mathbb{R}^{2\times n}$ is a two-row matrix where two rows find first-order discrete finite differences in horizontal and vertical directions of a pixel at the ith location, τ and μ are positive parameters for balancing between regularization and data fidelity terms. The main difficulty with the above problem is the non-differentiability of the TV and the ℓ_1-norm terms. RecPF solves this problem by reformulating it into a linearly constrained minimization problem by introducing two new auxiliary variables $\mathbf{w}$ and $\mathbf{z}$ to replace the TV and the ℓ_1-norm functions in (3.70), respectively. Thus, recasting the problem in (3.70) as

$$\begin{aligned} \hat{\mathbf{x}} = \ & \underset{\mathbf{w},\mathbf{z},\mathbf{x}}{\text{argmin}} \sum_i ||\mathbf{w}_i||_2 + \tau||\,\mathbf{z}||_1 + \frac{\mu}{2}||\mathbf{F}_u\mathbf{x} - \mathbf{y}||_2^2 \\ & \text{subject to } \ \mathbf{w}_i = \mathbf{D}_i\mathbf{x}, \quad \forall i; \\ & \qquad\qquad \mathbf{z} = \Psi\mathbf{x}, \end{aligned} \tag{3.71}$$

where $\mathbf{w} = [\mathbf{w}_1, \ldots, \mathbf{w}_n]$ and each $\mathbf{w}_i \in \mathbb{R}^2$; $\mathbf{z} \in \mathbb{R}^n$. The augmented Lagrangian function of the above problem can be defined as

$$L_\beta(\mathbf{w}, \mathbf{z}, \mathbf{x}, \boldsymbol{\lambda}_1, \boldsymbol{\lambda}_2) = \sum_i \phi_1(\mathbf{w}_i, \mathbf{D}_i\mathbf{x}, (\boldsymbol{\lambda}_1)_i) + \sum_i \phi_2(\mathbf{z}_i, \Psi_i\mathbf{x}, (\lambda_2)_i) + \frac{\mu}{2}||\mathbf{F}_u\mathbf{x} - \mathbf{y}||_2^2 \tag{3.72}$$

where

$$\phi_1(\mathbf{w}_i, \mathbf{D}_i\mathbf{x}, (\boldsymbol{\lambda}_1)_i) = ||\mathbf{w}_i||_2 - (\boldsymbol{\lambda}_1)_i{}^T(\mathbf{w}_i - \mathbf{D}_i\mathbf{x}) + \frac{\beta_1}{2}||\mathbf{w}_i - \mathbf{D}_i\mathbf{x}||_2^2,$$
$$\phi_2(z_i, \Psi_i\mathbf{x}, (\lambda_2)_i) = |z_i| - (\lambda_2)_i(z_i - \Psi_i\mathbf{x}) + \frac{\beta_2}{2}|z_i - \Psi_i\mathbf{x}|^2.$$

$\beta = \{\beta_1, \beta_2\}$, $\beta_1, \beta_2 > 0$ and $(\boldsymbol{\lambda}_1)_i \in \mathbb{R}^2$ and $(\lambda_2)_i \in \mathbb{R}, \forall_i$ are Lagrange multipliers. Now, steps of the ADMM algorithm may be applied iteratively until convergence to minimize $L_\beta(.)$ with respect to $\mathbf{w}$, $\mathbf{z}$ and $\mathbf{x}$ as follows:

STEP 1: Fix $\mathbf{x}$ and $\boldsymbol{\lambda}$ (Let, $\boldsymbol{\lambda} = (\boldsymbol{\lambda}_1, \boldsymbol{\lambda}_2)$) and minimize $L_\beta(\mathbf{w}, \mathbf{z}, \mathbf{x}, \boldsymbol{\lambda}_1, \boldsymbol{\lambda}_2)$ simultaneously with respect to $\mathbf{w}$ and $\mathbf{z}$.
Minimization with respect to $\mathbf{w}$

This can be done by solving a series of two-dimensional least square problems for all i as shown below:

$$\begin{aligned}
\mathbf{w}_i &= \underset{\mathbf{w}_i}{\text{minimize}}\, ||\mathbf{w}_i||_2 - (\lambda_1)_i^T(\mathbf{w}_i - \mathbf{D}_i\mathbf{x}) + \frac{\beta_1}{2}||\mathbf{w}_i - \mathbf{D}_i\mathbf{x}||_2^2 \\
&= \underset{\mathbf{w}_i}{\text{minimize}}\, ||\mathbf{w}_i||_2 + \frac{\beta_1}{2}||\mathbf{w}_i - \mathbf{D}_i\mathbf{x}||_2^2 - (\lambda_1)_i^T(\mathbf{w}_i - \mathbf{D}_i\mathbf{x}) \\
&= \underset{\mathbf{w}_i}{\text{minimize}}\, ||\mathbf{w}_i||_2 + \frac{\beta_1}{2}\Bigg(||\mathbf{w}_i - \mathbf{D}_i\mathbf{x}||_2^2 - \frac{(\lambda_1)_i^T}{\beta_1}(\mathbf{w}_i - \mathbf{D}_i\mathbf{x}) + \\
&\qquad \left(\frac{(\lambda_1)_i}{\beta_1}\right)^2 - \left(\frac{(\lambda_1)_i}{\beta_1}\right)^2\Bigg).
\end{aligned}$$

Neglecting the constant terms, we can write the above equation as

$$\mathbf{w}_i = \underset{\mathbf{w}_i}{\text{minimize}}\, ||\mathbf{w}_i||_2 + \frac{\beta_1}{2}\left(||\mathbf{w}_i - \mathbf{D}_i\mathbf{x} - \frac{(\lambda_1)_i}{\beta_1}||_2^2\right). \tag{3.73}$$

The solution of the above problem can be given by [55]

$$\mathbf{w}_i = \max\left\{\left\|\mathbf{D}_i\mathbf{x} + \frac{(\lambda_1)_i}{\beta_1}\right\|_2 - \frac{1}{\beta_1},\ 0\right\} \frac{\mathbf{D}_i\mathbf{x} + \frac{(\lambda_1)_i}{\beta_1}}{||\mathbf{D}_i\mathbf{x} + \frac{(\lambda_1)_i}{\beta_1}||_2}, \quad \forall i \tag{3.74}$$

Similarly,
Minimization with respect to $\mathbf{z}$

$$z_i = \underset{z_i}{\text{minimize}}\, |z_i| + \frac{\beta_2}{2}|z_i - \Psi_i\mathbf{x} - \frac{(\lambda_1)_i}{\beta_2}|^2, \qquad \text{(as done for } \mathbf{w}) \tag{3.75}$$

which leads to the component-wise solution given by

$$z_i = \max\left\{|\Psi_i\mathbf{x} + \tfrac{(\lambda_2)_i}{\beta_2}| - \tfrac{\tau}{\beta_2}, \quad 0\right\}\frac{\Psi_i\mathbf{x} + \frac{(\lambda_2)_i}{\beta_2}}{|\Psi_i\mathbf{x} + \frac{(\lambda_2)_i}{\beta_2}|}, \quad \forall i. \tag{3.76}$$

STEP 2: Results obtained for $\mathbf{w}$ and $\mathbf{z}$ in above with fixed $\boldsymbol{\lambda}$ are used for the minimization of L_β (.) with respect to $\mathbf{x}$.
Minimization with respect to $\mathbf{x}$

$$\underset{\mathbf{x}}{\text{minimize}} \; -\boldsymbol{\lambda}_1{}^T(\mathbf{w} - \mathbf{D}\mathbf{x}) + \tfrac{\beta_1}{2}||\mathbf{w} - \mathbf{D}\mathbf{x}||_2^2 - \boldsymbol{\lambda}_2{}^T(\mathbf{z} - \Psi\mathbf{x}) + \tfrac{\beta_2}{2}||\mathbf{z} - \Psi\mathbf{x}||_2^2 + \tfrac{\mu}{2}||\mathbf{F}_u\mathbf{x} - \mathbf{y}||_2^2,$$

where $\mathbf{w} = \begin{bmatrix}\mathbf{w}_1\\ \mathbf{w}_2\end{bmatrix}$, and $\mathbf{w}_j = [w_1^{(j)}, \ldots w_n^{(j)}]^T$ and $j \in \{1, 2\}$. Here, $\mathbf{D} \in \mathbb{R}^{2n \times n} = \begin{bmatrix}\mathbf{D}^{(1)}\\ \mathbf{D}^{(2)}\end{bmatrix}$, $\mathbf{D}^{(1)} \in \mathbb{R}^{n\times n}$ and $\mathbf{D}^{(2)} \in \mathbb{R}^{n\times n}$ being the two operators for finding first-order finite differences in horizontal and vertical directions of the image $\mathbf{x} \in \mathbb{R}^{n\times 1}$, respectively. Differentiating the above least square problem with respect to $\mathbf{x}$ transforms it into an equivalent set of normal equations is given by

$$\mathbf{M}\mathbf{x} = \mathbf{P}, \tag{3.77}$$

where

$$\mathbf{M} = \mathbf{D}^T\mathbf{D} + \mathbf{I} + \tfrac{\mu}{\beta}\mathbf{F}_u^T\mathbf{F}_u, \quad \text{and}$$
$$\mathbf{P} = \mathbf{D}^T\left(\mathbf{w} - \tfrac{\lambda_1^T}{\beta}\right) + \Psi^T\left(\mathbf{z} - \tfrac{\lambda_2^T}{\beta}\right) + \tfrac{\mu}{\beta}\mathbf{F}_u{}^T\mathbf{y},$$

for $\Psi^T\Psi = \mathbf{I}$ and $\beta_1 = \beta_2 = \beta$. Assuming periodic boundary conditions of $\mathbf{x}$, the matrix $\mathbf{D}^T\mathbf{D}$ can be diagonalized by 2-D FFT, $\mathbf{F}$ as $\mathbf{D}^{(1)}$ and $\mathbf{D}^{(2)}$ are block-circulant matrices. Applying FFT on both sides of the above equation, we get

$$\hat{\mathbf{M}}\mathbf{F}(\mathbf{x}) = \hat{\mathbf{P}}, \tag{3.78}$$

where

$$\hat{\mathbf{M}} = \hat{\mathbf{D}}^T\hat{\mathbf{D}} + \mathbf{I} + \tfrac{\mu}{\beta}\boldsymbol{\Phi}^T\boldsymbol{\Phi},$$
$$\hat{\mathbf{P}} = \mathbf{F}\left(\hat{\mathbf{D}}^T\mathbf{w} + \Psi^T\mathbf{z}\right) + \tfrac{\mu}{\beta}\boldsymbol{\Phi}^T\mathbf{y}$$
$$\hat{\mathbf{D}} = (\hat{\mathbf{D}}^{(1)}; \hat{\mathbf{D}}^{(2)}), \quad \text{and}$$
$$\hat{\mathbf{D}}^{(j)} = \mathbf{F}\,\mathbf{D}^{(j)}\,\mathbf{F}^T, \quad j = \{1, 2\}.$$

Since $\hat{\mathbf{M}}$ is a diagonal matrix, we can easily obtain $\mathbf{F}(\mathbf{x})$ from the above ecquision and then by inverse FFT we can get the solution $\mathbf{x}$. Thus, one can minimize L_β with respect to $(\mathbf{w}, \mathbf{z}, \mathbf{x})$ by applying Eqs. 3.74, 3.76 and 3.78 iteratively followed by immediate updating of $\boldsymbol{\lambda}$ until convergence. We summarize the whole algorithm in Algorithm 10. According to [58, Theorem 2.1], the algorithm converges for any $\beta > 0$ and $\gamma \in \left(0, \left(\sqrt{5}+1\right)/2\right)$ from an arbitrary starting point.

Algorithm 10 Reconstruction from Partial Fourier Data (RecPF)

Input: $\mathbf{F}_u$, $\mathbf{y}$, Ψ, τ, $(\mu, \beta_1, \beta_2) > 0$, and $\gamma \in \left(0, \left(\sqrt{5}+1\right)/2\right)$

Initialization: $\mathbf{x}^{(0)} \leftarrow \mathbf{F}_u^T\mathbf{y}$, $\boldsymbol{\lambda}_1{}^{(0)}$, $\boldsymbol{\lambda}_2{}^{(0)}$ and k $\leftarrow 0$

1: **while** not converged **do**

2: $\quad \mathbf{w}_i{}^{(k+1)} \leftarrow \max\left\{\|\mathbf{D}_i\mathbf{x}^{(k)} + \frac{\left(\boldsymbol{\lambda}_1^{(k)}\right)_i}{\beta_1}\|_2 - \frac{1}{\beta_1},\ 0\right\} \cdot \frac{\mathbf{D}_i\mathbf{x}^{(k)} + \frac{\left(\boldsymbol{\lambda}_1^{(k)}\right)_i}{\beta_1}}{\|\mathbf{D}_i\mathbf{x}^{(k)} + \frac{\left(\boldsymbol{\lambda}_1^{(k)}\right)_i}{\beta_1}\|_2}, \quad \forall i$

3: $\quad z_i^{(k+1)} \leftarrow \max\left\{|\Psi_i\mathbf{x}^{(k)} + \frac{\left(\boldsymbol{\lambda}_2^{(k)}\right)_i}{\beta_2}| - \frac{\tau}{\beta_2},\ 0\right\} \cdot \frac{\Psi_i\mathbf{x}^{(k)} + \frac{\left(\boldsymbol{\lambda}_2^{(k)}\right)_i}{\beta_2}}{|\Psi_i\mathbf{x}^{(k)} + \frac{\left(\boldsymbol{\lambda}_2^{(k)}\right)_i}{\beta_2}|}, \quad \forall i$

4: $\quad \mathbf{x}^{(k+1)} \leftarrow \mathbf{F}^{-1}\left(\frac{\hat{\mathbf{P}}^{(k+1)}}{\hat{\mathbf{M}}}\right)$

5: $\quad (\boldsymbol{\lambda}_1)_i{}^{(k+1)} \leftarrow (\boldsymbol{\lambda}_1)_i{}^{(k)} - \gamma\beta_1\left(\mathbf{w}_i{}^{(k+1)} - \mathbf{D}_i\mathbf{x}^{(k+1)}\right), \quad \forall i$

6: $\quad (\boldsymbol{\lambda}_2)_i{}^{(k+1)} \leftarrow (\boldsymbol{\lambda}_2)_i{}^{(k)} - \gamma\beta_2\left(z_i{}^{(k+1)} - \Psi_i\mathbf{x}^{(k+1)}\right), \quad \forall i$

7: $\quad k \leftarrow k+1$

8: **end while**

Output: $\mathbf{x}^* \leftarrow \mathbf{x}^{(k)}$

3.3.3 Algorithm Based on Bregman Iteration

Bregman Distance

Bregman iteration can be used to solve a wide range of convex optimization problems [14]. Osher et al. [48] first time applied it to the *Rudin–Osher–Fatemi* (ROF) model for denoising [52]. Also, it is applied for solving compressed sensing problems with ℓ_1 minimization in [59]. The core term involved in the Bregman iteration is the *"Bregman distance"* which we define here for its importance in our study. Consider, a convex function $f(\mathbf{x})$. The Bregman distance associated with this function between two points $\mathbf{x}$ and $\mathbf{v}$ can be written as

$$D_g^{\mathbf{p}}(\mathbf{x}, \mathbf{v}) = f(\mathbf{x}) - f(\mathbf{v}) - \langle \mathbf{p},\ \mathbf{x} - \mathbf{v} \rangle, \tag{3.79}$$

where $\mathbf{p}$ is in the subgradient of f at $\mathbf{v}$. In other words, the Bregman distance $D_f^{\mathbf{p}}(\mathbf{x}, \mathbf{v})$ is the difference between the value of f at x and the first-order Taylor series approximation of f around the point v.

Consider now a constrained optimization problem:

$$\begin{aligned} \underset{\mathbf{x}}{\text{minimize}} \quad & f(\mathbf{x}) \\ \text{subject to} \quad & \mathbf{A}\mathbf{x} = \mathbf{y}. \end{aligned} \tag{3.80}$$

Using the penalty function method the equivalent unconstrained problem can be defined as

$$\underset{\mathbf{x}}{\text{minimize}} \quad f(\mathbf{x}) + \tfrac{\lambda}{2}||\mathbf{A}\mathbf{x} - \mathbf{y}||_2^2, \tag{3.81}$$

where λ is the weight of the penalty function. If $f(\mathbf{x}) = ||\Psi \mathbf{x}||_1$, the above problem is a basis pursuit (BP) problem. The Bregman iteration steps to minimize the above problem iteratively can be written as [14]

$$\begin{aligned} \mathbf{x}^{(k+1)} &= \underset{\mathbf{x}}{\text{minimize}} \quad D_f^{\mathbf{p}}\left(\mathbf{x}, \mathbf{x}^{(k)}\right) + \tfrac{\lambda}{2}||\mathbf{A}\mathbf{x} - \mathbf{y}||_2^2 \\ &= \underset{\mathbf{x}}{\text{minimize}} \quad f(\mathbf{x}) - \left\langle \mathbf{p}^{(k)}, \ \mathbf{x} - \mathbf{x}^{(k)} \right\rangle + \tfrac{\lambda}{2}||\mathbf{A}\mathbf{x} - \mathbf{y}||_2^2 \end{aligned} \tag{3.82}$$

$$\mathbf{p}^{(k+1)} = \mathbf{p}^{(k)} - \lambda \mathbf{A}^T \left(\mathbf{A}\mathbf{x}^{(k+1)} - \mathbf{y}\right). \tag{3.83}$$

Assuming that the iterative solution $\mathbf{x}^{(k)}$ of Eq. 3.82 satisfies $\mathbf{A}\mathbf{x}^{(k)} = \mathbf{y}$ then $\mathbf{x}^{(k)}$ also converges to the optimal solution $\mathbf{x}_{\text{opt}}$ of the basis pursuit problem in (3.81) according to [59, see Theorem 3.2] which may be seen from the following analysis. We know that $D_g^{\mathbf{p}}(\mathbf{x}, \mathbf{v}) \geq 0$, therefore, for any $\mathbf{x}$ we can write

$$\begin{aligned} f\left(\mathbf{x}^{(k)}\right) &\leqslant f(\mathbf{x}) - \left\langle \mathbf{x} - \mathbf{x}^{(k)}, \ \mathbf{p}^{(k)} \right\rangle \\ &= f(\mathbf{x}) - \left\langle \mathbf{x} - \mathbf{x}^{(k)}, \ \mathbf{A}^T \left(\mathbf{y}^{(k)} - \mathbf{A}\mathbf{x}^{(k)}\right) \right\rangle \\ &= f(\mathbf{x}) - \left\langle \mathbf{A}\mathbf{x} - \mathbf{A}\mathbf{x}^{(k)}, \ \mathbf{y}^{(k)} - \mathbf{A}\mathbf{x}^{(k)} \right\rangle \\ &= f(\mathbf{x}) - \left\langle \mathbf{A}\mathbf{x} - \mathbf{y}, \ \mathbf{y}^{(k)} - \mathbf{y} \right\rangle. \end{aligned} \tag{3.84}$$

From above, we find that during any Bregman iteration its corresponding solution $\mathbf{x}^{(k)}$ would satisfy $f\left(\mathbf{x}^{(k)}\right) \leqslant f(\mathbf{x})$, i.e., the Bregman iteration converges if and only if any $\mathbf{x}_{\text{opt}}$ satisfies $\mathbf{A}\mathbf{x}_{\text{opt}} = \mathbf{y}$. Thus, solutions obtained from the Bregman iteration and that of the BP problem are the same on convergence. Now we will simplify the Bregman iteration in Eq. 3.82 according to the analysis given in [59, Sect. 3].

At $k = 0$, $\mathbf{p}^{(1)} = \mathbf{p}^{(0)} - \lambda \mathbf{A}^T \left(\mathbf{A}\mathbf{x}^{(1)} - \mathbf{y}\right)$. Assuming that, $\mathbf{p}^{(0)} = \mathbf{0}$ and $\mathbf{y}^{(1)} = \mathbf{y}$ implies $\mathbf{p}^{(1)} = \lambda \mathbf{A}^T \left(\mathbf{y}^{(1)} - \mathbf{A}\mathbf{x}^{(1)}\right)$. With the above assumption, we may rewrite the steps of the Bregman iteration as follows:

$$
\begin{aligned}
\mathbf{x}^{(k+1)} &= \underset{\mathbf{x}}{\text{minimize}} \;\; D_f^p\left(\mathbf{x}, \mathbf{x}^{(k)}\right) + \tfrac{\lambda}{2}||\mathbf{Ax} - \mathbf{y}||_2^2 \\
&= \underset{\mathbf{x}}{\text{minimize}} \;\; f(\mathbf{x}) - \left\langle \mathbf{p}^{(k)},\; \mathbf{x} \right\rangle + \tfrac{\lambda}{2}||\mathbf{Ax} - \mathbf{y}||_2^2 + c_1 \\
&= \underset{\mathbf{x}}{\text{minimize}} \;\; f(\mathbf{x}) + \tfrac{\lambda}{2}\left(||\mathbf{Ax} - \mathbf{y}||_2^2 - 2\left\langle \mathbf{y}^{(k)} - \mathbf{Ax}^{(k)}, \mathbf{Ax} \right\rangle\right) + c_2 \\
&= \underset{\mathbf{x}}{\text{minimize}} \;\; f(\mathbf{x}) + \tfrac{\lambda}{2}\left(||\mathbf{Ax} - \mathbf{y}||_2^2 - 2\left\langle \mathbf{Ax} - \mathbf{y}, \mathbf{y}^{(k)} - \mathbf{Ax}^{(k)} \right\rangle + \right. \\
&\qquad\qquad \left. ||\mathbf{y}^{(k)} - \mathbf{Ax}^{(k)}||_2^2\right) + c_3 \\
&= \underset{\mathbf{x}}{\text{minimize}} \;\; f(\mathbf{x}) + \tfrac{\lambda}{2}||\mathbf{Ax} - \mathbf{y} - \mathbf{y}^{(k)} + \mathbf{Ax}^{(k)}||_2^2 + c_3 \\
&= \underset{\mathbf{x}}{\text{minimize}} \;\; f(\mathbf{x}) + \tfrac{\lambda}{2}||\mathbf{Ax} - \mathbf{y}^{(k)}||_2^2 + c_3 \qquad (3.85) \\
\mathbf{y}^{(k+1)} &= \mathbf{y} + \mathbf{y}^{(k)} - \mathbf{Ax}^{(k)}, \qquad (3.86)
\end{aligned}
$$

where c_1, c_2 and c_3 denote terms independent of $\mathbf{x}$. The second expression is obtained by using values of $p^{(k)}$ and $p^{(k+1)}$ with their assumptions done above and substituting them in Eq. 3.83. It indicates that the error (difference between observed and estimated data) of the previous iteration is added to the current iteration to speed up the convergence. In the following, we discuss about the *split-Bregman algorithm* based on the above concepts.

3.3.3.1 Split-Bregman Algorithm

Goldsten and Osher proposed the *split-Bregman* algorithm based on Bregman iterations [33]. It can efficiently solve the TV-ℓ_1-ℓ_2 model of CS-MRI which we rewrite below for further analysis.

$$\hat{\mathbf{x}} = \underset{\mathbf{x}}{\text{argmin}} \frac{\mu}{2}||\mathbf{F}_u\mathbf{x} - \mathbf{y}||_2^2 + ||\Psi\mathbf{x}||_1 \; + \; ||\mathbf{x}||_{TV}, \qquad (3.87)$$

where $||\mathbf{x}||_{TV} = \sum_i \sqrt{(\mathbf{D}^{(1)}\,\mathbf{x})_i^2 + (\mathbf{D}^{(2)}\,\mathbf{x})_i^2}$ for isotropic TV.

Following [33, Sect. 4.2], we solve the above minimization problem by using dual concepts of variable splitting and Bregman iteration as follows. First, apply variable splitting to the above minimization problem by introducing new variables: $\mathbf{z} = \Psi\mathbf{x}$, $\omega_1 = \mathbf{D}^{(1)}\mathbf{x}$ and $\omega_2 = \mathbf{D}^{(2)}\mathbf{x}$. Next, decompose the main problem after variable substitutions into a series of independent subproblems containing just one of the three new variables and then followed by their alternate minimizations by Bregman iterations. Thus, we obtain

$$
\begin{aligned}
&\underset{\mathbf{x},\omega_1,\omega_2,\mathbf{z}}{\text{argmin}} \frac{\mu}{2}||\mathbf{F}_u\mathbf{x} - \mathbf{y}||_2^2 + ||\mathbf{z}||_1 + ||(\omega_1, \omega_2)||_2 + \frac{\lambda}{2}||\omega_1 - \mathbf{D}^{(1)}\,\mathbf{x} - \mathbf{y}_1||_2^2 + \\
&\frac{\lambda}{2}||\omega_2 - \mathbf{D}^{(2)}\,\mathbf{x} - \mathbf{y}_2||_2^2 + \frac{\gamma}{2}||\mathbf{z} - \Psi\mathbf{x} - \mathbf{y}_3||_2^2
\end{aligned}
\qquad (3.88)
$$

where $||(\omega_1, \omega_2)||_2 = \sum_i \sqrt{\left|\omega_{1,i}\right|^2 + \left|\omega_{2,i}\right|^2}$; $\mathbf{y}_1$, $\mathbf{y}_2$, and $\mathbf{y}_3$ are chosen for updating a Bregman iteration.

Then, minimizing with respect to $\mathbf{z}$, ω_1 and ω_2, we obtain the following pairs of intermediate solutions and Bregman updates as follows:
Minimization with respect to $\mathbf{z}$

$$\begin{aligned}\mathbf{z}^{(k+1)} &= \mathscr{T}\left(\Psi \mathbf{x}^{(k+1)} + \mathbf{y}_3^{(k)}, \tfrac{1}{\gamma}\right) \\ \mathbf{y}_3^{(k+1)} &= y_3^{(k)} + \left(\Psi \mathbf{x}^{(k+1)} - \mathbf{z}^{(k+1)}\right).\end{aligned}$$

Minimization with respect to ω_1

$$\begin{aligned}\omega_1{}^{(k+1)} &= \max(\mathbf{p}^{(k)} - \tfrac{1}{\lambda}, 0)\frac{\mathbf{D}^{(1)}\mathbf{x}^{(k)} + \mathbf{y}_1^{(k)}}{\mathbf{p}^{(k)}} \\ \mathbf{y}_1{}^{(k+1)} &= \mathbf{y}_1{}^{(k)} + \left(\mathbf{D}^{(1)}\mathbf{x}^{(k+1)} - \omega_1{}^{(k+1)}\right).\end{aligned}$$

and
Minimization with respect to ω_2

$$\begin{aligned}\omega_2{}^{(k+1)} &= \max(\mathbf{p}^{(k)} - \tfrac{1}{\lambda}, 0)\frac{\mathbf{D}^{(2)}\mathbf{x}^{(k)} + \mathbf{y}_2{}^{(k)}}{\mathbf{p}^{(k)}} \\ \mathbf{y}_2{}^{(k+1)} &= \mathbf{y}_2{}^{(k)} + \left(\mathbf{D}^{(2)}\mathbf{x}^{(k+1)} - \omega_2{}^{(k+1)}.\right),\end{aligned}$$

where $p_i^{(k)} = \sqrt{|(\mathbf{D}^{(1)}\mathbf{x}^{(k)})_i + y_{1\,i}{}^{(k)}|^2 + |(\mathbf{D}^{(2)}\mathbf{x}^{(k)})_i + y_{2\,i}{}^{(k)}|^2}$ is obtained by a technique reported in [33, Sect. 4, Eq. 4.6] and [53]. After getting solutions of $\mathbf{z}^{(k)}$, $\omega_1^{(k)}$ and $\omega_2^{(k)}$ in the current iteration, we substitute them into Eq. 3.88 for its minimization with respect to $\mathbf{x}$. Thus, the final subproblem for minimization is

$$\begin{aligned}\mathbf{x}^{k+1} = \underset{\mathbf{x}}{\operatorname{argmin}}\ & \tfrac{\mu}{2}||\mathbf{F}_u\mathbf{x} - \mathbf{y}||_2^2 + \tfrac{\lambda}{2}||\omega_1^{(k)} - \mathbf{D}^{(1)}\,\mathbf{x} - \mathbf{y}_1^{(k)}||_2^2 + \\ & \tfrac{\lambda}{2}||\omega_2^{(k)} - \mathbf{D}^{(2)}\,\mathbf{x} - \mathbf{y}_2^{(k)}||_2^2 + \tfrac{\gamma_1}{2}||\mathbf{z}^{(k)} - \Psi\mathbf{x} - \mathbf{y}_3^{(k)}||_2^2,\end{aligned} \tag{3.89}$$

which is quadratic in $\mathbf{x}$, so it is differentiable with respect to $\mathbf{x}$. Thus, differentiating the above equation we obtain:

$$\left(\mu\,\mathbf{F}_u^T\mathbf{F}_u + \lambda \mathbf{D}^{(1)^T}\mathbf{D}^{(1)} + \lambda \mathbf{D}^{(2)^T}\mathbf{D}^{(2)} + \gamma \Psi^T \Psi\right)\mathbf{x}^{(k+1)} = \mathbf{M}^{(k)}, \tag{3.90}$$

where

$$\mathbf{M}^{(k)} = \mu\,\mathbf{F}_u^T\mathbf{y} + \lambda\mathbf{D}^{(1)^T}\left(\omega_1^{(k)} - \mathbf{y}_1^{(k)}\right) + \\ \lambda\mathbf{D}^{(2)^T}\left(\omega_2^{(k)} - \mathbf{y}_2^{(k)}\right) + \gamma\Psi^T\left(\mathbf{z}^{(k)} - \mathbf{y}_3^{(k)}\right)$$

Due to the orthogonal property of wavelet and Fourier transforms, one can write $\Psi^T\Psi = \mathbf{F}^T\mathbf{F} = \mathbf{I}$, and also assuming $\mathbf{D} = \begin{bmatrix}\mathbf{D}^{(1)} \\ \mathbf{D}^{(2)}\end{bmatrix}$. So, we get

$$\left(\mu\,\mathbf{F}_u^T\mathbf{F}_u + \lambda\mathbf{D}^T\mathbf{D} + \gamma I\right)\mathbf{x}^{(k+1)} = \mathbf{M}^{(k)} \tag{3.91}$$

For obtaining the solution for $\mathbf{x}^{(k+1)}$, the matrix in the left-hand side has to be inverted. This can be done easily by diagonalizing this matrix using the FFT operator, $\mathbf{F}$ as follows:

$$\begin{aligned}
\mathbf{F}^T\mathbf{F}\left(\mu\,\mathbf{F}_u^T\mathbf{F}_u + \lambda\mathbf{D}^T\mathbf{D} + \gamma I\right)\mathbf{F}^T\mathbf{F}\mathbf{x}^{(k+1)} &= \mathbf{M}^{(k)} \\
\mathbf{F}^T\left(\mu\,\Phi^T\Phi + \lambda\hat{\mathbf{D}}^T\hat{\mathbf{D}} + \gamma I\right)\mathbf{F}\mathbf{x}^{(k+1)} &= \mathbf{M}^{(k)} \\
\mathbf{F}^T\mathbf{K}\mathbf{F}\mathbf{x}^{(k+1)} &= \mathbf{M}^{(k)} \\
\mathbf{x}^{(k+1)} &= \mathbf{F}\mathbf{K}^{-1}\mathbf{F}^T\mathbf{M}^{(k)},
\end{aligned}$$

where $\hat{\mathbf{D}} = \mathbf{F}\,\mathbf{D}\,\mathbf{F}^T$ and $\mathbf{K} = \left(\mu\Phi^T\Phi + \lambda\hat{\mathbf{D}}^T\hat{\mathbf{D}} + \gamma\mathbf{I}\right)$.

The algorithmic details are summarized in Algorithm 11. The main advantage of this method is the achievement of faster convergence of the ℓ_1 regularized problems due to iterative updates of the true signal approximation error. Also, the algorithm makes extensive use of the Gauss–Seidel and Fourier transform methods which can be parallelized very easily. These advantages make the algorithm most suitable for large-scale problems.

3.4 Composite Splitting

There is yet another relatively new class of splitting algorithms known as the *composite splitting* which covers the ideas of both operator and variable splitting. This class of algorithms are particularly suitable for composite regularization problems like the TV-ℓ_1-ℓ_2 problem [38, 39]. The main idea of this algorithm is first split the composite problem into different simpler subproblems using the idea of variable splitting. Next, solve each subproblem independently by efficient operator splitting technique and finally linearly combine solutions of individual subproblems to get the solution of the composite problem. Let us consider a general minimization problem as

$$\underset{\mathbf{x}}{\text{minimize}}\, f(\mathbf{x}) = h(\mathbf{x}) + \sum_{j=1}^{p} \lambda_j\, g_j(\mathbf{x}), \tag{3.92}$$

Algorithm 11 Split Bregman Algorithm

Input: $\mathbf{F}_u, \mathbf{y}, \Psi$

Initialization: $\mathbf{x}^{(0)} \leftarrow \mathbf{F}_u^T \mathbf{y},\ ({\omega_1}^{(0)},\ {\omega_2}^{(0)},\ \mathbf{z}^{(0)},\ {\mathbf{y}_1}^{(0)},\ {\mathbf{y}_2}^{(0)},\ \mathbf{y}_3^{(0)}) \leftarrow \mathbf{0},\ k \leftarrow 0$

while not converge **do**
    for i = 1: N **do**
        $\mathbf{x}^{(k+1)} \leftarrow \mathbf{F}\mathbf{K}^{-1}\mathbf{F}^T\mathbf{M}^{(k)}$
        ${\omega_1}^{(k+1)} \leftarrow \max(\mathbf{p}^{(k)} - \frac{1}{\lambda}, 0)\frac{\mathbf{D}^{(1)}\mathbf{x}^{(k)} + {\mathbf{y}_1}^{(k)}}{\mathbf{p}^{(k)}}$
        ${\omega_2}^{(k+1)} \leftarrow \max(\mathbf{p}^{(k)} - \frac{1}{\lambda}, 0)\frac{\mathbf{D}^{(2)}\mathbf{x}^{(k)} + {\mathbf{y}_2}^{(k)}}{\mathbf{p}^{(k)}}$
        $\mathbf{z}^{(k+1)} \leftarrow \mathscr{T}\left(\left(\Psi\mathbf{x}^{(k+1)}\right) + \mathbf{y}_3^{(k)},\ \frac{1}{\gamma}\right)$
        ${\mathbf{y}_1}^{(k+1)} \leftarrow {\mathbf{y}_1}^{(k)} + \left(\mathbf{D}^{(1)}\mathbf{x}^{(k+1)} - {\omega_1}^{(k+1)}\right)$
        ${\mathbf{y}_2}^{(k+1)} \leftarrow {\mathbf{y}_2}^{(k)} + \left(\mathbf{D}^{(2)}\mathbf{x}^{(k+1)} - {\omega_2}^{(k+1)}\right)$
        $\mathbf{y}_3^{(k+1)} \leftarrow \mathbf{y}_3^{(k)} + \left(\Psi\mathbf{x}^{(k+1)} - \mathbf{z}^{(k+1)}\right)$
    end for
    $\mathbf{y}^{(k+1)} \leftarrow \mathbf{y}^{(k)} + \mathbf{y} - \mathbf{F}_u\mathbf{x}^{(k+1)}$
end while

Output: $\mathbf{x}^* \leftarrow \mathbf{x}^{(k+1)}$

where $g_j\,(.)$ are non-smooth convex functions, $h\,(\mathbf{x}) = \frac{1}{2}||\mathbf{A}\mathbf{x} - \mathbf{y}||_2^2$ is a continuously differentiable function with Lipschitz constant L_h. If $p = 1$, the above problem can be solved very easily by using an operator splitting technique. However, for $p > 1$, above minimization problem will have multiple regularization terms. For example, let us consider $p = 2$ and $g_1(\mathbf{x}) = ||(x)||_1, g_2(\mathbf{x}) = ||\mathbf{x}||_{TV}$. Then, solving it by the operator splitting technique would become highly expensive computationally. Huang et al. [38] proposed an efficient algorithm known as the *composite splitting denoising* (CSD) based on the idea of composite splitting, that is, dividing the minimization problem into p simpler subproblems and then linearly combining their solutions to arrive at the desired solution of the composite problem.

3.4.1 Composite Splitting Denoising

Authors in [38], considered the above problem as a denoising problem. They solved it by using the concept of "composite splitting", i.e., (1) First split the variable $\mathbf{x}$ into multiple variables, i.e., $\{\mathbf{x}_j\}_{j=1,2,\ldots,p}$ to generate different subproblems; (2) Apply operator splitting to solve the subproblems independently with respect to each of $\mathbf{x}_j$; (3) Obtain the solution $\mathbf{x}$ by a linear combination of all $\mathbf{x}_j s$ obtained above. They termed this algorithm as the *composite splitting denoising* (CSD) algorithm. The convergence of the algorithm is based on the proofs given in [18, Th. 3.4] and [38, Th. 3.1].

According to these theorems, consider that each subproblem has a unique global minimum which belongs to a particular set $\mathbf{G}$, i.e., $\{\mathbf{x}_j\}_{j=1,2,\ldots,p} \in \mathbf{G}$ which are linearly combined to get the target solution $\mathbf{x}^{(k)}$ at the kth iteration. Then, the sequence

$\{\mathbf{x}^{(k)}\}\ k \in \mathbb{N}$ would converge weakly to a point in **G** iteratively under the following conditions [17, 18]:

1. $\lim_{\|\mathbf{x}\| \to +\infty} \{f_1(\mathbf{x}) + \cdots + f_p(\mathbf{x})\} = +\infty$. This means that $\mathbf{G} \neq \emptyset$ (see proof in [18, Prop. 3.2]) and
2. $(\mathbf{0}, \ldots, \mathbf{0}) \in \text{sri}\{(\mathbf{x} - \mathbf{x}_1, \ldots, \mathbf{x} - \mathbf{x}_p) \mid \mathbf{x} \in \mathscr{H}, \mathbf{x}_1 \in \text{dom } f_1, \ldots, \mathbf{x}_p \in \text{dom } f_p\}$ where 'sri' refers to strong relative interior and $\mathscr{H}$ is a real Hilbert space (see [18, Props. 3.2 and 3.3]). This implies that $\text{dom } (f(\mathbf{x}_1) + \ldots + f(\mathbf{x}_p)) = \text{dom } (f(\mathbf{x}_1)) \cap \ldots \cap \text{dom } (f(\mathbf{x}_p)) \neq \emptyset$.

Steps of the CSD algorithm are outlined in Algorithm 12. In the algorithm, auxiliary vectors $\{\mathbf{z}_j\}_{j=1,\ldots,p}$ are used for faster convergence. For each subproblem, its solution at the kth iteration $\mathbf{x}_j^{(k)}$ is subtracted from the composite solution $\mathbf{x}^{(k)}$ and the error is added to the auxiliary vector $\mathbf{z}_j$ of the previous iteration, i.e., $\mathbf{z}_j^{(k-1)}$. This improves the convergence of the main problem. Another important feature of this algorithm is that both shrinkage operations and updating of auxiliary variables are carried out simultaneously indicating parallel structure of the algorithm.

In the following, we also discuss further improvements of the CSD algorithm by combining it with the iterative shrinkage algorithms. This development has led to the introduction of two new algorithms, namely, the *composite splitting Algorithm* (CSA) and the *Fast*-CSA as reported in [38]. The corresponding algorithms are given in Algorithms 13 and 14 and demonstrate very good performance in MR image reconstruction [39].

3.4.2 Composite Splitting Algorithm (CSA)

CSA is the combination of the CSD and the IST algorithm. In [39], the authors minimize the TV-ℓ_1-ℓ_2 model of the CS-MRI using the CSA, i.e.,

Algorithm 12 Composite Splitting Denoising (CSD)

Input: $\rho \leftarrow L_h$, $\{\lambda_j\}_{j=1,\ldots,p}$, $\{\mathbf{z}_j^{(0)}\}_{j=1,\ldots,p} \leftarrow \mathbf{A}^T\mathbf{y}$

Initialization: $t^{(1)}$, $k \leftarrow 0$

while not converged **do**
  for j = 1: p **do**
    $\mathbf{x}_j \leftarrow \mathscr{T}\left(\mathbf{z}_j^{(k)}, \frac{\lambda_j}{\rho}\right)$
  end for
  $\mathbf{x}^{(k+1)} \leftarrow \frac{1}{p}\sum_{j=1}^{p} \mathbf{x}_j$
  for j = 1: p **do**
    $\mathbf{z}_j^{(k+1)} \leftarrow \mathbf{z}_j^{(k)} + \mathbf{x}^{(k+1)} - \mathbf{x}_j$
  end for
  $k \leftarrow k + 1$
end while

Output: $\mathbf{x}^* \leftarrow \mathbf{x}^{(k)}$

$$\hat{\mathbf{x}} = \underset{\mathbf{x}}{\text{argmin}}\ \tfrac{1}{2}||\mathbf{F}_u\mathbf{x} - \mathbf{y}||_2^2 + \lambda_1||\,\Psi\mathbf{x}||_1 \ + \lambda_2\ ||\mathbf{x}||_{TV} \tag{3.93}$$

Using the concept of composite splitting, we decompose the above problem into two subproblems, one is the ℓ_1-regularization subproblem and other is the TV regularization subproblem as given below

$$\hat{\mathbf{x}}_1 = \underset{\mathbf{x}}{\text{argmin}}\ \tfrac{1}{2}||\mathbf{F}_u\mathbf{x} - \mathbf{y}||_2^2 + \lambda_1||\,\Psi\mathbf{x}||_1 \tag{3.94}$$

$$\hat{\mathbf{x}}_2 = \underset{\mathbf{x}}{\text{argmin}}\ \tfrac{1}{2}||\mathbf{F}_u\mathbf{x} - \mathbf{y}||_2^2 + \lambda_2\ ||\mathbf{x}||_{TV}. \tag{3.95}$$

The IST algorithm can easily solve the ℓ_1-regularization subproblem (i.e., the $\ell_1 - \ell_2$ problem) using soft-thresholding. On the other hand, the TV regularization subproblem (i.e., the TV $- \ell_2$ problem) is solved by using a dual approach of discrete TV regularization proposed in [3]. Assuming solutions of these individual subproblems by $\mathbf{x}_1$ and $\mathbf{x}_2$, respectively, the final solution $\mathbf{x}$ of the TV-ℓ_1-ℓ_2 model is then obtained by simply averaging $\mathbf{x}_1$ and $\mathbf{x}_2$. The steps of the CSA is summarized in Algorithm 13. In step 6 of the algorithm, the *project* function is defined as

$$\mathbf{x}_i = \text{project}\,(\mathbf{x}_i,\ [l,\ u]) = \begin{cases} \mathbf{x}_i, & \text{if } l \le \mathbf{x}_i \le u \\ l, & \text{if } \mathbf{x}_i < l \\ u, & if\ \mathbf{x}_i > u, \end{cases}\ , \tag{3.96}$$

where i represents the pixel locations of the image $\mathbf{x}$; l and u denotes the range of pixels of MR images. For example, $l = 0$ and $u = 255$ for 8-bit gray scale MR images.

Algorithm 13 Composite Splitting Algorithm (CSA)

Input: $\mathbf{F}_u,\ \mathbf{y},\ \Psi,\ \rho \leftarrow L,$
Initialization: $\mathbf{x}^{(0)} \leftarrow \mathbf{F}_u^T\mathbf{y},\ \mathbf{r}^{(1)} \leftarrow \mathbf{x}^{(0)},\ \lambda_1,\ \lambda_2,\ k \leftarrow 1$
1: **while** not converged **do**
2: $\quad \mathbf{x}_g \leftarrow \mathbf{r}^{(k)} - \rho\nabla f\left(\mathbf{r}^{(k)}\right)$
3: $\quad \mathbf{x}_1 \leftarrow \mathscr{T}\left(\mathbf{x}_g,\ \frac{\lambda_1}{\rho}\right)$
4: $\quad \mathbf{x}_2 \leftarrow \mathscr{T}\left(\mathbf{x}_g,\ \frac{\lambda_2}{\rho}\right)$
5: $\quad \mathbf{x}^{(k)} \leftarrow (\mathbf{x}_1 + \mathbf{x}_2)/2$
6: $\quad \mathbf{x}^{(k)} \leftarrow \text{project}\left(\mathbf{x}^{(k)},\ [l,\ u]\right)$
7: $\quad \mathbf{r}^{(k+1)} \leftarrow \mathbf{x}^{(k)}$
8: $\quad k \leftarrow k + 1$
9: **end while**
Output: $\mathbf{x}^* \leftarrow \mathbf{x}^{(k)}$

3.4.3 Fast Composite Splitting Algorithm (FCSA)

Similarly, another faster version of the CSA, i.e., the FCSA is based on the combination of the CSD and the FISTA [4]. The steps of the FCSA are summarized in Algorithm 14. Recently, a similar algorithm is proposed in [22], where the authors combine the CSD and the ALM algorithms to solve the TV-ℓ_1-ℓ_2 model. It has been observed that the algorithm gives better reconstruction and faster convergence than the FCSA. More details about the experimental results of the above algorithms are discussed in Chap. 5.

Algorithm 14 Fast Composite Splitting Algorithm (FCSA)

Input: $\mathbf{F}_u,\ \mathbf{y},\ \Psi,\ \rho \leftarrow L,$
Initialization: $\mathbf{x}^{(0)} \leftarrow \mathbf{F}_u^T\mathbf{y},\ \mathbf{r}^{(1)} \leftarrow \mathbf{x}^{(0)},\ \lambda_1,\ \lambda_2,\ k \leftarrow 1$
1: **while** not converged **do**
2: $\quad \mathbf{x}_g \leftarrow \mathbf{r}^{(k)} - \rho\nabla f\left(\mathbf{r}^{(k)}\right)$
3: $\quad \mathbf{x}_1 \leftarrow \mathscr{T}\left(\mathbf{x}_g,\ \frac{\lambda_1}{\rho}\right)$
4: $\quad \mathbf{x}_2 \leftarrow \mathscr{T}\left(\mathbf{x}_g,\ \frac{\lambda_2}{\rho}\right)$
5: $\quad \mathbf{x}^{(k)} \leftarrow (\mathbf{x}_1 + \mathbf{x}_2)/2$
6: $\quad \mathbf{x}^{(k)} \leftarrow \text{project}\left(\mathbf{x}^{(k)},\ [l,\ u]\right)$
7: $\quad t^{(k+1)} \leftarrow \frac{\left(1+\sqrt{1+4(t^{(k)})^2}\right)}{2}$
8: $\quad \mathbf{r}^{(k+1)} \leftarrow \mathbf{x}^{(k)} + \frac{t^{(k)}-1}{t^{(k+1)}}\left(\mathbf{x}^{(k)} - \mathbf{x}^{(k-1)}\right)$
9: $\quad k \leftarrow k+1$
10: **end while**
Output: $\mathbf{x}^* \leftarrow \mathbf{x}^{(k)}$

3.5 Non-splitting Method

This class of algorithms belongs to the traditional methods which are applied in CS-MRI but directly do not fall under any of the spitting categories mentioned in this paper. A few very popular schemes in the non-splitting category include the *nonlinear conjugate gradient* (NCG) method, the *gradient projection for sparse reconstruction* (GPSR) and the *truncated Newton interior-point method* (TNIPM). The main limitation of these algorithms is the slow convergence. However, some of these algorithms can produce results comparable to those of operator and variable splitting algorithms. But they perform poorly with respect to the composite splitting algorithms. Although they are not targeted to achieve fast CS-MRI reconstruction, we discuss them in the following to complete the discussion on important developments of CS-MRI algorithms.

3.5.1 Nonlinear Conjugate Gradient Method

Before going to discuss the *nonlinear conjugate gradient* (NCG) method, we start with the background of the *conjugate gradient* (CG) method. The CG method was originally developed by Hestenes and Stiefel in 1952 [37] based on the concept of deriving the optimal solution to a system of linear equations as the linear combination of a set of conjugate directions. To define conjugate directions, we say that a pair of nonzero vectors $\mathbf{u}_1$ and $\mathbf{u}_2$ are conjugate with respect to $\mathbf{A}$ if the inner product

$$\langle \mathbf{u}_1, \ \mathbf{A}\mathbf{u}_2 \rangle = {\mathbf{u}_1}^T \mathbf{A}\mathbf{u}_2 = 0.$$

Now, let us consider the system:

$$\mathbf{A}\mathbf{x} = \mathbf{y}, \tag{3.97}$$

where $\mathbf{A} \in \mathbb{R}^{(n \times n)}$ is assumed to be symmetric, i.e., $\mathbf{A} = \mathbf{A}^T$, positive definite, i.e., $\mathbf{x}^T\mathbf{A}\mathbf{x} > 0$, and real. The unique solution of the system $\mathbf{x}^*$ may be obtained by solving a quadratic function of the form $f(\mathbf{x}) = \frac{1}{2}\mathbf{x}^T\mathbf{A}\mathbf{x} - \mathbf{y}^T\mathbf{x}$. If n is very large, the direct method would take too much time to give the unique minimizer $\mathbf{x}^*$. Instead, we may solve it iteratively by approximating a solution $\mathbf{x}^{(k)}$ at each iteration. In this method, we start with a initial guess $\mathbf{x}^{(0)}$ and at every iteration we must select a solution which would be closer approximation to the final solution $\mathbf{x}^*$. To start with, a search direction is selected in the negative gradient direction of $f(\mathbf{x})$ at $\mathbf{x} = \mathbf{x}^{(0)}$, i.e., $p^{(0)} = \mathbf{y} - \mathbf{A}\mathbf{x}^{(0)}$. Then, subsequent directions $\{\mathbf{p}^{(k)}\}$ are selected such that they are conjugate to the gradient. If we define the residue at the kth step: $\mathbf{r}^{(k)} = \mathbf{y} - \mathbf{A}\mathbf{x}^{(k)}$ then the next search direction may be obtained from the current residue $\mathbf{r}^{(k)}$ and all previous search directions. Since, the next search directions are also conjugate to each other, the following formula is defined for finding next search directions $\{\mathbf{p}^{(k)}\}$ using the Gram–Schmidt Orthonormalization procedure:

$$\mathbf{p}^{(k)} = \mathbf{r}^{(k)} - \sum_{i<k} \frac{\mathbf{p}^{(i)T}\mathbf{A}\mathbf{r}^{(k)}}{\mathbf{p}^{(i)T}\mathbf{A}\mathbf{p}^{(i)}}\mathbf{p}^{(i)} \tag{3.98}$$

$$= \mathbf{r}^{(k)} + \sum_{i<k} \gamma^{(i)}\mathbf{p}^{(i)}, \tag{3.99}$$

where $\gamma^{(i)} = -\frac{\mathbf{p}^{(i)T}\mathbf{A}\mathbf{r}^{(k)}}{\mathbf{p}^{(i)T}\mathbf{A}\mathbf{p}^{(i)}}$, $\forall i < k$. After getting the new search directions, the next iteration for the optimal solution may be defined as:

$$\mathbf{x}^{(k+1)} = \mathbf{x}^{(k)} + t^{(k)}\mathbf{p}^{(k)}, \tag{3.100}$$

where $t^{(k)} = \frac{\langle \mathbf{p}^{(k)}, \mathbf{r}^{(k-1)} \rangle}{\langle \mathbf{p}^{(k)}, A\mathbf{p}^{(k)} \rangle} = \frac{\langle \mathbf{r}^{(k)}, \mathbf{r}^{(k)} \rangle}{\langle \mathbf{p}^{(k)}, \mathbf{A}\mathbf{p}^{(k)} \rangle}$. We summarize the iterative CG method in Algorithm 15.

Algorithm 15 Iterative Conjugate Gradient Algorithm

Input: $\mathbf{y}$, $\mathbf{A}$

Initialization: $k \leftarrow 0$, $\mathbf{x}^{(0)} \leftarrow \mathbf{A}^T\mathbf{y}$, $\mathbf{r}^{(0)} \leftarrow \mathbf{y} - \mathbf{A}\mathbf{x}^{(0)}$, and $\mathbf{p}^{(0)} \leftarrow \mathbf{r}^{(0)}$

1: **while** not converged **do**

2: $t^{(k)} \leftarrow \frac{|\mathbf{r}^{(k)}|^2}{\langle \mathbf{p}^{(k)}, \mathbf{A}\mathbf{p}^{(k)} \rangle}$

3: $\mathbf{x}^{(k+1)} \leftarrow \mathbf{x}^{(k)} + t^{(k)}\mathbf{p}^{(k)}$

4: $\mathbf{r}^{(k+1)} \leftarrow \mathbf{r}^{(k)} - t^{(k)}\mathbf{A}\mathbf{p}^{(k)}$

5: $\gamma^{(k)} \leftarrow \frac{|\mathbf{r}^{(k+1)}|^2}{|\mathbf{r}^{(k)}|^2}$

6: $\mathbf{p}^{(k+1)} \leftarrow \mathbf{r}^{(k+1)} + \gamma^{(k)}\mathbf{p}^{(k)}$

7: $k \leftarrow k + 1$

8: **end while**

Output: $\mathbf{x}^* \leftarrow \mathbf{x}^k$

The nonlinear conjugate gradient (NCG) method is the generalized version of the CG method. It is also an iterative optimization algorithm suitable for large-scale sparse data. Lustig et al. in [42] solved the CS-MRI problem P_1 in Eq. 2.5 using the NCG algorithm. In particular, they solve the following nonlinear minimization problem using the iterative CG method:

$$\hat{\mathbf{x}} = \underset{\mathbf{x}}{\operatorname{argmin}}\, f(\mathbf{x}) = \frac{1}{2}||\mathbf{F}_u\mathbf{x} - \mathbf{y}||_2^2 + \lambda||\,\Psi\mathbf{x}||_1 \tag{3.101}$$

For the above problem, gradient of the cost function $f(\mathbf{x})$ can be written as

$$\nabla f(\mathbf{x}) = \mathbf{F}_u^T(\mathbf{F}_u\mathbf{x} - \mathbf{y}) + \lambda\nabla||\Psi\mathbf{x}||_1, \tag{3.102}$$

where the ℓ_1 norm is a non-smooth function, therefore not directly differentiable. However, in their work, Lustig et al. [42] did an approximation to the absolute value $|\Psi\mathbf{x}|$ to simplify the ℓ_1-norm term as

$$|\Psi\mathbf{x}| \approx \sqrt{(\Psi\mathbf{x})^T(\Psi\mathbf{x}) + \mu},$$

where μ is a positive smoothing parameter, $\mu \in \left[10^{-15}, 10^{-6}\right]$. With this approximation,

$$\frac{d|\,(\Psi\mathbf{x})\,|}{d\,(\Psi x)} \approx \frac{\Psi\mathbf{x}}{\sqrt{(\Psi\mathbf{x})^T(\Psi\mathbf{x}) + \mu}}.$$

Therefore, the gradient of $f(\mathbf{x})$ can be approximated as

$$\nabla f(\mathbf{x}) \approx \mathbf{F}_u^T(\mathbf{F}_u\mathbf{x} - \mathbf{y}) + \lambda\Psi^T\mathbf{W}^{-1}\Psi\mathbf{x}, \tag{3.103}$$

where $\mathbf{W} = \text{Diag}\left(\{w_i\}_{i=1,\ldots,n}\right)$ with $w_i = \sqrt{(\Psi\mathbf{x})_i^T(\Psi\mathbf{x})_i + \mu}$.

After the gradient is computed approximately, next requirement is the computation of the step size with which each iteration approaches the final optimal solution $\mathbf{x}^*$. To find the suitable step size in the descent direction, Lustig et al. adopted the backtracking line search technique [13, Algo. 9.2]. The steps of the nonlinear conjugate gradient algorithm for CS-MRI are summarized in Algorithm 16. In the above algorithm, ξ and β are the line search parameters (default values are: $\xi = 0.05$, $\beta = 0.6$). The number of iterations required for convergence of the NCG varies depending on size and nature of the data. For example, it took about 150 iterations to reconstruct a 480×92 angiogram image at fivefold accelerations using the NCG method.

Algorithm 16 Nonlinear Conjugate Gradient Algorithm

Input: $\mathbf{y}$, $\mathbf{F}_u$, Ψ, λ, ξ and β
Initialization: $k \leftarrow 0$, $\mathbf{x}^{(0)} \leftarrow \mathbf{F}_u^T \mathbf{y}$, $\mathbf{g}^{(0)} \leftarrow \nabla f\left(\mathbf{x}^{(0)}\right)$, $\Delta\mathbf{x}^{(0)} \leftarrow -\mathbf{g}^{(0)}$
while not converged **do**
  $t \leftarrow 1$
  while $\left(f\left(\mathbf{x}^{(k)} + t\Delta\mathbf{x}^{(k)}\right) > f\left(\mathbf{x}^{(k)}\right) + \xi t.\ \mathrm{Re}\left(\mathbf{g}^{(k)^T} \Delta\mathbf{x}^{(k)}\right)\right)$ **do**
    $t \leftarrow \beta t$
  end while
  $\mathbf{x}^{(k+1)} \leftarrow \mathbf{x}^{(k)} + t\Delta\mathbf{x}^{(k)}$
  $\mathbf{g}^{(k+1)} \leftarrow \nabla f\left(\mathbf{x}^{(k+1)}\right)$
  $\gamma^{(k)} \leftarrow \frac{||\mathbf{g}^{(k+1)}||_2^2}{||\mathbf{g}^{(k)}||_2^2}$
  $\Delta\mathbf{x}^{(k+1)} \leftarrow -\mathbf{g}^{(k+1)} + \gamma^{(k)} \Delta\mathbf{x}^{(k)}$
  $k \leftarrow k+1$
end while
Output: $\mathbf{x}^* \leftarrow \mathbf{x}^{(k)}$

3.5.2 *Gradient Projection for Sparse Reconstruction*

In this method, first, we formulate an unconstrained problem with an ℓ_1 penalty term into a bound-constrained quadratic program (BCQP) by convexification of the non-differentiable ℓ_1 penalty term. Let us consider the following minimization problem:

$$\arg\min_{\mathbf{x}} \frac{1}{2} ||\mathbf{y} - \mathbf{A}\mathbf{x}||_2^2 + \lambda ||\mathbf{x}||_1. \tag{3.104}$$

As stated above, we first the above as a BCQP by splitting $\mathbf{x}$ into its positive and negative parts [29], i.e.,

$$\mathbf{x} = \mathbf{v}_1 - \mathbf{v}_2, \quad \mathbf{v}_1 \geq 0, \quad \mathbf{v}_2 \geq 0.$$

where $v_{1i} = (x_i)_+$, $v_{2i} = (-x_i)_+$ for all $i = 1, 2, \ldots, n$, and $(a)_+ = \max\{0, a\}$. Thus, $||\mathbf{x}||_1 = \mathbf{1}_n^T \mathbf{v}_1 + \mathbf{1}_n^T \mathbf{v}_2$, where $\mathbf{1}_n^T$ is a unit vector of length n.

Above simplification of the ℓ_1 term transforms the minimization problem as the bound-constrained quadratic program (BCQP):

$$\begin{aligned} \min_{\mathbf{v}_1, \mathbf{v}_2} \quad & \tfrac{1}{2}||\mathbf{y} - \mathbf{A}(\mathbf{v}_1 - \mathbf{v}_2)||_2^2 + \lambda \mathbf{1}_n^T \mathbf{v}_1 + \lambda \mathbf{1}_n^T \mathbf{v}_2 \\ \text{subject to} \quad & \mathbf{v}_1 \geq 0, \quad \mathbf{v}_2 \geq 0 \end{aligned} \tag{3.105}$$

.

This formulation is valid even if $v_{1i} = 0$ or $v_{2i} = 0$ for $i = 1, 2, \ldots, n$. This is because if we add a nonnegative shift parameter $\mathbf{s}$ to $\mathbf{v}_1$ and $\mathbf{v}_2$ such that $\mathbf{v}_1 \leftarrow (\mathbf{v}_1 + s)$ and $\mathbf{v}_2 \leftarrow (\mathbf{v}_2 + s)$ then the ℓ_2 term is unaffected but increases the remaining two terms by $2\tau \mathbf{1}_n^T \mathbf{s}$. In the following, we write the following standard BCQP form:

$$\begin{aligned} \min_{z} \quad & \mathbf{c}^T \mathbf{z} + \tfrac{1}{2}\mathbf{z}^T \mathbf{B} \mathbf{z} \equiv f(\mathbf{z}) \\ \text{subject to} \quad & \mathbf{z} \geq 0 \end{aligned} \tag{3.106}$$

,

where $\mathbf{z} = \begin{bmatrix} \mathbf{v}_1 \\ \mathbf{v}_2 \end{bmatrix}$, $\mathbf{b} = \mathbf{A}^T \mathbf{y}$, $\mathbf{c} = \lambda \mathbf{1}_{2n} + \begin{bmatrix} -\mathbf{b} \\ \mathbf{b} \end{bmatrix}$, and $\mathbf{B} = \begin{bmatrix} \mathbf{A}^T\mathbf{A} & -\mathbf{A}^T\mathbf{A} \\ -\mathbf{A}^T\mathbf{A} & \mathbf{A}^T\mathbf{A} \end{bmatrix}$

The fundamental idea of the GPSR algorithm is to carry out the following two steps until convergence:

1. ***Projection***:
   Search $\mathbf{z}^{(k)}$ in each iteration along the negative gradient: $-\nabla f\left(\mathbf{z}^{(k)}\right)$. Then, project the the result $\left(\mathbf{z}^{(k)} - t^{(k)} \nabla f\left(\mathbf{z}^{(k)}\right)\right)$ onto the feasible set to obtain $\mathbf{w}^{(k)}$. Thus, $\mathbf{w}^{(k)} = \left(\mathbf{z}^{(k)} - t^{(k)} \nabla f\left(\mathbf{z}^{(k)}\right)\right)_+$, where $(.)_+$ represents the projection operation.
2. ***Line search***:
   Now, take a step along the feasible direction $(\mathbf{w}^{(k)} - \mathbf{z}^{(k)})$ using a step size $\gamma^{(k)}$. That is $\mathbf{z}^{(k+1)} = \mathbf{z}^{(k)} + \gamma^{(k)}\left(\mathbf{w}^{(k)} - \mathbf{z}^{(k)}\right)$.

Depending on the techniques for the estimation of $t^{(k)}$ and $\gamma^{(k)}$, the GPSR algorithm may be further divided into two categories. One is the *GPSR-basic* and other is the *Barzilai–Borwein GPSR* (GPSR-BB). Detailed theories involved in these algorithms may be obtained from [27] and papers referred therein. In the following, we present them very briefly for our analysis only.

3.5.2.1 GPSR-Basic

In the basic version of the GPSR algorithm, it is ensured that the objective function f decreases at every iteration. That is in each iteration, we search the next iterate along the steepest descent direction, i.e., $-\nabla f\left(\mathbf{z}^{(k)}\right)$ and assume that the step $\left(\mathbf{z}^{(k)} - t^{(k)} \nabla f\left(\mathbf{z}^{(k)}\right)\right)$ is within the feasible set, i.e.,

$$\mathbf{z}^{(k+1)} = \left(\mathbf{z}^{(k)} - t^{(k)} \nabla f\left(\mathbf{z}^{(k)}\right)\right)_{+}. \tag{3.107}$$

Then, we perform the backtracking line search to find the suitable step size $t^{(k)}$ in each iteration as given below

$$\begin{aligned}
&\text{while}\\
&\left(f\left(\mathbf{z}^{(k+1)}\right) > f\left(\mathbf{z}^{(k)}\right) - \mu \nabla f\left(\mathbf{z}^{(k)}\right)^{T}\left(\mathbf{z}^{(k)} - \mathbf{z}^{(k+1)}\right)\right)\\
&\qquad t^{(k)} = \beta t^{(k)}\\
&\qquad \mathbf{z}^{(k+1)} = \left(\mathbf{z}^{(k)} - t^{(k)} \nabla f\left(\mathbf{z}^{(k)}\right)\right)_{+}\\
&\text{endwhile}
\end{aligned}$$

where μ and β are backtracking parameters. At each iteration, we choose t_0 as the solution of

$$t_0 = \arg\min_{t} f\left(\mathbf{z}^{(k)} - t\mathbf{g}^{(k)}\right), \tag{3.108}$$

where $\mathbf{g}_i^{(k)} = \begin{cases} \left(\nabla f\left(\mathbf{z}^{(k)}\right)\right)_i, & \text{if } \mathbf{z}_i^{(k)} > 0 \text{ or } \left(\nabla f\left(\mathbf{z}^{(k)}\right)\right)_i < 0 \\ 0, & \text{otherwise.} \end{cases}$

This leads to the closed-form formula for t_0 as [27]

$$t_0 = \frac{\left(\mathbf{g}^{(k)}\right)^{T} \mathbf{g}^{(k)}}{\left(\mathbf{g}^{(k)}\right)^{T} \mathbf{B}\mathbf{g}^{(k)}}.$$

In every iteration, it is confirmed that the value of t_0 lies within the interval $[t_{\min}, t_{\max}]$ by selecting $t_0 = \text{mid}\,(t_{\min}, t_0, t_{\max})$, so that t_0 is not too large or too small.

Algorithm 17 GPSR-basic

Input: $\mathbf{y}, \mathbf{A}$
Initialization: $\mathbf{z}^{(0)}$, $\beta \in (0, 1)$, $\mu \in \left(0, 1/2\right)$ and $k \leftarrow 0$
1: **while** not converge **do**
2: $t_0 \leftarrow \frac{\left(\mathbf{g}^{(k)}\right)^{T} \mathbf{g}^{(k)}}{\left(\mathbf{g}^{(k)}\right)^{T} \mathbf{B}\mathbf{g}^{(k)}}$
3: $t^{(k)} \leftarrow \text{mid}\,(t_{\min}, t_0, t_{\max})$
4: **while** $\left(f\left(\mathbf{z}^{(k+1)}\right) > f\left(\mathbf{z}^{(k)}\right) - \mu \nabla f\left(\mathbf{z}^{(k)}\right)^{T}\left(\mathbf{z}^{(k)} - \mathbf{z}^{(k+1)}\right)\right)$ **do**
5: $t^{(k)} \leftarrow \beta t^{(k)}$
6: $\mathbf{z}^{(k+1)} \leftarrow \left(\mathbf{z}^{(k)} - t^{(k)} \nabla \mathbf{F}\left(\mathbf{z}^{(k)}\right)\right)_{+}$
7: **end while**
8: $k \leftarrow k + 1$
9: **end while**
Output: $\mathbf{z}^{*} \leftarrow \mathbf{z}^{(k)}$

3.5.2.2 GPSR-BB

In contrast to the GPSR-basic, the GPSR-BB does not guarantee that the objective function $f(\mathbf{z})$ would decrease at each iteration. It was originally applied in the context of unconstrained minimization of a smooth nonlinear function $f(\mathbf{z})$. It computes a step $\boldsymbol{\delta}^{(k)} = -\left(\mathbf{H}^{(k)}\right)^{-1} \nabla f\left(\mathbf{z}^{(k)}\right)$, where $\mathbf{H}^{(k)}$ is an approximation to the Hessian matrix of $f\left(\mathbf{z}^{(k)}\right)$. Then, $\mathbf{H}^{(k)}$ is estimated by the simple formula: $\mathbf{H}^{(k)} = \eta^{(k)}\mathbf{I}$, where $\eta^{(k)}$ is chosen in such a way that it satisfies the Lipschitz condition in the least square sense:

$$\nabla f\left(\mathbf{z}^{(k)}\right) - \nabla f\left(\mathbf{z}^{(k-1)}\right) \approx \eta^{(k)}\left(\mathbf{z}^{(k)} - \mathbf{z}^{(k-1)}\right). \tag{3.109}$$

Finally, the solution is updated by the following formula:

$$\mathbf{z}^{(k+1)} = \mathbf{z}^{(k)} - \left(\eta^{(k)}\right)^{-1} \nabla f\left(\mathbf{z}^{(k)}\right). \tag{3.110}$$

The BB approach discussed above is extended for solving the BCQP. Here, the GPSR steps discussed in Sect. 3.5.2 would be carried out as follows:

First, compute the direction:

$$\boldsymbol{\delta}^{(k)} = \left(\mathbf{z}^{(k)} - t^{(k)}\nabla f\left(\mathbf{z}^{(k)}\right)\right)_{+} - \mathbf{z}^{(k)}, \tag{3.111}$$

where $t^k = \left(\eta^{(k)}\right)^{-1}$ and it is restricted in the interval $[t_{\min}, t_{\max}]$. It is computed by using the Barzilai–Borwein spectral rule [2]:

$$t^{(k)} = \begin{cases} \text{mid}\left\{t_{\min}, \frac{\|\boldsymbol{\delta}^{(k)}\|_2^2}{\upsilon^{(k)}}, t_{\max}\right\}, & if \;\; \upsilon^{(k)} \neq 0 \\ t_{\max}, & if \;\; \upsilon^{(k)} = 0 \end{cases}, \tag{3.112}$$

where $\upsilon^{(k)} = \left(\boldsymbol{\delta}^{(k)}\right)^T \mathbf{B}\, \boldsymbol{\delta}^{(k)}$.

Next, apply line search to compute:

$$\mathbf{z}^{(k+1)} = \mathbf{z}^{(k)} + \gamma^{(k)}\boldsymbol{\delta}^{(k)} \tag{3.113}$$

where parameter $\gamma^{(k)}$ is chosen in the interval $\gamma^{(k)} \in [0, 1]$ such that $f\left(\mathbf{z}^{(k)} + \gamma^{(k)}\,\delta^{(k)}\right)$ is minimized. A simple closed-form expression for $\gamma^{(k)}$ can be obtained by using [27]

$$\gamma^{(k)} = \text{mid}\left\{0, \frac{\left(\boldsymbol{\delta}^{(k)}\right)^T \nabla f\left(\mathbf{z}^{(k)}\right)}{\left(\boldsymbol{\delta}^{(k)}\right)^T \mathbf{B}\, \boldsymbol{\delta}^{(k)}}, 1\right\}.$$

In the above, parameter $\gamma^{(k)}$ assumed unity if $\left(\boldsymbol{\delta}^{(k)}\right)^T \mathbf{B}\, \boldsymbol{\delta}^{(k)} = 0$. The parameter $\gamma^{(k)}$ computed by the above approach also removes the property of the Barzilai–Borwein approximation, i.e., chances of increasing the objective function in certain iteration which on the other hand improves the overall performances.

The algorithmic steps of the GPSR-basic and the GPSR-BB are summarized in Algorithms 17 and 18, respectively. Each iteration of the GPSR involves matrix-vector multiplications of $\mathbf{A}$ and $\mathbf{A}^T$ together with a few inner products of vectors of length n. The GPSR algorithm performs well for large-scale problem. These algorithms converge at least five times faster than the ISTA.

For solving the CS-MRI reconstruction problem using the PCG algorithm, the measurement matrix $\mathbf{A}$ should be selected as the product of an $m \times n$ binary matrix $\boldsymbol{\Phi}$ with an $n \times n$ DFT matrix $\mathbf{F}$. Undersampling matrix $\boldsymbol{\Phi}$ is formed by randomly picking m rows of an $n \times n$ Identity matrix. Different acceleration or scan time reduction factors may be achieved by setting suitable value of m. Since, MR image is compressible in transform domain we need to replace $||\mathbf{x}||_1$ by $||\boldsymbol{\Psi}\mathbf{x}||_1$ in Eq. 3.104, where $\boldsymbol{\Psi}$ is the wavelet basis.

Algorithm 18 GPSR-BB

Input: $\mathbf{y}, \mathbf{A}$

Initialization: $\mathbf{z}^{(0)}, t_{\min}, t_{\max}, t^{(0)} \in [t_{\min}, t_{\max}]$ and $k \leftarrow 0$

1: **while** not converge **do**

2: $\boldsymbol{\delta}^{(k)} = \left(\mathbf{z}^{(k)} - t^{(k)}\nabla f\left(\mathbf{z}^{(k)}\right)\right)_+ - \mathbf{z}^{(k)}$

3: $\gamma^{(k)} = \text{mid}\left\{0, \dfrac{\left(\boldsymbol{\delta}^{(k)}\right)^T \nabla f(\mathbf{z}^{(k)})}{\left(\boldsymbol{\delta}^{(k)}\right)^T \mathbf{B}\, \boldsymbol{\delta}^{(k)}}, 1\right\}$

4: $\mathbf{z}^{(k+1)} = \mathbf{z}^{(k)} + \gamma^{(k)}\boldsymbol{\delta}^{(k)}$

5: $\upsilon^{(k)} = \left(\boldsymbol{\delta}^{(k)}\right)^T \mathbf{B}\, \boldsymbol{\delta}^{(k)}$

6: $t^{(k)} = \begin{cases} \text{mid}\left\{t_{\min}, \dfrac{\left\|\boldsymbol{\delta}^{(k)}\right\|_2^2}{\upsilon^{(k)}}, t_{\max}\right\}, & if\ \upsilon^{(k)} \neq 0 \\ t_{\max}, & if\ \upsilon^{(k)} = 0 \end{cases}$

7: $k \leftarrow k+1$

8: **end while**

Output: $\mathbf{z}^* \leftarrow \mathbf{z}^{(k)}$

3.5.3 *Truncated Newton Interior-Point Method*

In the *truncated Newton interior-point method* (TNIPM), we first formulate the ℓ_1 regularized least squares problem (LSP) (cf. P_1 in Eq. 2) into a convex quadratic problem using linear inequality constraints [40, Sec. IV]. For a general setting, let us consider the following ℓ_1 regularized LSP:

$$\arg\min_{\mathbf{x}} \ ||\mathbf{A}\mathbf{x} - \mathbf{y}||_2^2 + \lambda||\mathbf{x}||_1, \tag{3.114}$$

where $\mathbf{x} \in \mathbb{R}^n$ is a sparse signal, $\mathbf{y} \in \mathbb{R}^m$ is measured data, $\mathbf{A} \in \mathbb{R}^{m\times n}$ and $m \ll n$. The above problem can be rewritten as a convex quadratic problem (QP) with inequality constraints as

$$\begin{aligned} &\arg\min_{\mathbf{x}} \ ||\mathbf{Ax}-\mathbf{y}||_2^2+\lambda\sum_{i=1}^{n}u_i \\ &\text{subject to } -u_i \leqslant x_i \leqslant u_i, \qquad i=1,\ldots,n. \end{aligned} \tag{3.115}$$

where $\mathbf{u} \in \mathbb{R}^n$. Let us now define the logarithmic barrier function for constraints $-u_i \leqslant x_i \leqslant u_i$ as [13, Chap. 11]:

$$Q(\mathbf{x},\mathbf{u}) = -\frac{1}{t}\sum_{i=1}^{n}\log(u_i+x_i)-\frac{1}{t}\sum_{i=1}^{n}\log(u_i-x_i), \tag{3.116}$$

where $t > 0$. The inequality constrained problem in Eq. 3.115 can be approximately converted into an equality constrained problem with the help of the logarithmic barrier function in Eq. 3.116, so that Newton's method may be applied. The central path of the equivalent unconstrained problem consists of the solution of

$$\arg\min_{\mathbf{x},\mathbf{u}} \phi_t(\mathbf{x},\mathbf{u}) = \arg\min_{\mathbf{x},\mathbf{u}} \ t\,||\mathbf{Ax}-\mathbf{y}||_2^2+t\sum_{i=1}^{n}\lambda\, u_i+Q(\mathbf{x},\mathbf{u}) \tag{3.117}$$

where t varies from 0 to ∞. The associated central path contains the unique minimizer $(\mathbf{x}^*(t), \mathbf{u}^*(t))$ of the convex function $\phi_t(\mathbf{x},\mathbf{u})$. Generally, $\phi_t(\mathbf{x},\mathbf{u})$ is $\frac{2n}{t}$ suboptimal therefore the central path leads to the optimal solution. In the interior-point method (IPM), a sequence of points on the central path are computed with the increasing values of t. The process is terminated when we reach $\frac{2n}{t} \leq \varepsilon$, where ε is the target duality gap [13, Ch.11]. In this method, $\phi_t(\mathbf{x},\mathbf{u})$ is minimized using the Newton's method where the search direction $(\Delta\mathbf{x}, \Delta\mathbf{u})$ is obtained using the following Newton's system:

$$\mathbf{H}\begin{bmatrix}\Delta\mathbf{x}\\ \Delta\mathbf{u}\end{bmatrix} = -\mathbf{g}, \tag{3.118}$$

where $\mathbf{H} = \nabla^2\phi_t(\mathbf{x},\mathbf{u}) \in \mathbb{R}^{2n\times 2n}$ is the Hessian and $\mathbf{g} = \nabla\phi_t(\mathbf{x},\mathbf{u}) \in \mathbb{R}^{2n}$ is the gradient of $\phi_t(\mathbf{x},\mathbf{u})$ at $(\mathbf{x},\mathbf{u})$. For large-scale problems, solving the above system accurately is computationally prohibitive. Therefore, $(\Delta\mathbf{x}, \Delta\mathbf{u})$ is computed approximately by iteratively solving a sequence of conjugate gradient (CG) steps [7]. However, if $\mathbf{H}$ is not well conditioned then the CG algorithm would converge very slowly. Due to this fact, the preconditioned CG (PCG) [23, Sec. 6.6] steps are used for faster convergence of the CG algorithm. A good preconditioner dramatically improves the convergence of the CG algorithm as it reduces the condition number of the matrix $\mathbf{H}$. With the inclusion of the PCG method in the traditional IPM for solving the Newton's system iteratively, a modified algorithm has been developed in [40] known by the name the *truncated Newton interior-point method* (TNIPM). In this algorithm, the Hessian $\mathbf{H}$ may be expressed in a compact form as shown below

$$\mathbf{H} = t\nabla^2 ||\mathbf{A}\mathbf{x} - \mathbf{y}||_2^2 + \nabla^2 Q(\mathbf{x}, \mathbf{u}) = \begin{bmatrix} 2t\mathbf{A}^T\mathbf{A} + \Lambda_1 & \Lambda_2 \\ \Lambda_2 & \Lambda_1 \end{bmatrix}, \tag{3.119}$$

where

$$\Lambda_1 = \text{diag}\left(\frac{2(u_1^2+x_1^2)}{(u_1^2-x_1^2)^2}, \ldots, \frac{2(u_n^2+x_n^2)}{(u_n^2-x_n^2)^2}\right) \in \mathbb{R}^n,$$

$$\Lambda_2 = \text{diag}\left(\frac{-4u_1^2x_1^2}{(u_1^2-x_1^2)^2}, \ldots, \frac{-4u_n^2x_n^2}{(u_n^2-x_n^2)^2}\right) \in \mathbb{R}^n,$$

here diag (.) denote the diagonal matrix. Similarly, the gradient $\mathbf{g}$ may be written as

$$\mathbf{g} = \begin{bmatrix} \mathbf{g}_1 \\ \mathbf{g}_2 \end{bmatrix} \in \mathbb{R}^{2n}, \tag{3.120}$$

where

$$\mathbf{g}_1 = \nabla_{\mathbf{x}} \ \phi_t(\mathbf{x}, \mathbf{u}) = 2t\mathbf{A}^T(\mathbf{A}\mathbf{x} - \mathbf{y}) + \begin{bmatrix} \frac{2x_1}{(u_1^2-x_1^2)} \\ \vdots \\ \frac{2x_n}{(u_n^2-x_n^2)} \end{bmatrix} \in \mathbb{R}^n,$$

$$\mathbf{g}_2 = \nabla_{\mathbf{u}} \ \phi_t(\mathbf{x}, \mathbf{u}) = 2t\lambda\mathbf{1} - \begin{bmatrix} \frac{2u_1}{(u_1^2-x_1^2)} \\ \vdots \\ \frac{2u_n}{(u_n^2-x_n^2)} \end{bmatrix} \in \mathbb{R}^n.$$

The PCG algorithm solves the Newton's system in Eq. 3.118 using a symmetric and positive definite preconditioner $\mathbf{P} \in \mathbb{R}^{2n\times 2n}$ as given below

$$\mathbf{P} = \begin{bmatrix} 2\tau t\mathbf{I} + \Lambda_1 & \Lambda_2 \\ \Lambda_2 & \Lambda_1 \end{bmatrix}, \tag{3.121}$$

where τ is a positive constant. The above approximation works very well when variations in the diagonal elements of $\mathbf{A}^T\mathbf{A}$ are not very high. An iteration of the PCG algorithm involves some inner products and some matrix-vector multiplications. The number of iterations required depends on the value of the regularization parameter λ and the stopping criterion. Extensive simulations show that for large-scale problems, several hundreds of PCG iterations are required for the convergence [40, Sect. IV(C)].

In [40, Sect. V(B)], the authors demonstrated the implementation of CS-MRI reconstruction problem using the TNIPM. For that, they acquire 205 phase encode lines randomly out of 512 phase encode lines to achieve an acceleration factor of 2.5. In simulation, this is done by removing some lines from the 2D DFT matrix, called as the partial Fourier matrix. Then, this is multiplied by the image to get the undersam-

Algorithm 19 Truncated newton IPM for ℓ_1 regularized LSPs

Input: $\mathbf{y}$, $\mathbf{A}$

Initialization: $\mathbf{x}^{(0)} \leftarrow \mathbf{0}$, $\mathbf{u}^{(0)} \leftarrow \mathbf{1}$, $\beta \in (0, 1)$, $\mu \in (0, 0.5)$, and $k \leftarrow 0$

1: **while** not converge **do**
2: Compute $(\Delta\mathbf{x}^{(k)}, \Delta\mathbf{u}^{(k)})$
3: **while** $\left(f\left(\mathbf{x}^{(k)} + \Delta\mathbf{x}^{(k)}\right) > f\left(\mathbf{x}^{(k)}\right) + \mu\, s\, \nabla f\left(\mathbf{x}^{(k)}\right)^T \Delta\mathbf{x}^{(k)}\right)$ **do**
4: $s \leftarrow s\,\beta$
5: **end while**
6: $(\mathbf{x}^{(k+1)}, \mathbf{u}^{(k+1)}) \leftarrow (\mathbf{x}^{(k)}, \mathbf{u}^{(k)}) + s(\Delta\mathbf{x}^{(k)}, \Delta\mathbf{u}^{(k)})$
7: $\nu^{(k+1)} \leftarrow 2\,s\left(\mathbf{A}\mathbf{x}^{(k)} - \mathbf{y}\right)$
8: Compute duality gap η
9: $k \leftarrow k + 1$
10: **end while**

Output: $\mathbf{z}^* \leftarrow \mathbf{z}^{(k)}$

pled k-space data. As MR images are sparse in the wavelet domain, Daubechies-4 wavelet transform is used as a sparsifying transform. Reconstructed results from the TNIPM are then compared with the zero-filled linear reconstruction results using the inverse Fourier transform. Results show that artifacts are significantly less in the case of TNIPM as compared to the later method. We summarize the main steps of the TNIPM in Algorithm 19.

3.6 Conclusions

This chapter presents a brief review of fast ℓ_1-minimization-based CS-MRI reconstruction algorithms with mathematical details. Different ℓ_1-minimization-based CS-MRI reconstruction algorithms are categorized into four groups and detailed their mathematical analysis with relative advantages and disadvantages. From the analysis, we conclude that the composite splitting-based reconstruction algorithms would give better reconstruction as compared to others. This is because the composite splitting-based algorithms utilize features of both the operator and variable splitting algorithms. Detailed experimental results with real and synthetic data sets are demonstrated in the next chapter.

References

1. Afonso, M., Bioucas-Dias, J., Figueiredo, M.: Fast image recovery using variable splitting and constrained optimization. IEEE Trans. Image Process. **19**(9), 2345–2356 (2010)
2. Barzilai, J., Borwein, J.M.: Two-point step size gradient methods. IMA J. Numer. Anal. **8**, 141–148 (1988)
3. Beck, A., Teboulle, M.: Fast gradient-based algorithms for constrained total variation image denoising and deblurring problems. IEEE Trans. Image Process. **18**(11), 2419–2434 (2009)

4. Beck, A., Teboulle, M.: A fast iterative shrinkage-thresholding algorithm for linear inverse problems. SIAM J. Imaging Sci. **2**(1), 183–202 (2009)
5. Becker, S., Bobin, J., Candes, E.J.: NESTA: a fast and accurate first-order method for sparse recovery. SIAM J. Imaging Sci. **4**(1), 1–39 (2011)
6. van den Berg, E., Friedlander, M.P.: Probing the pareto frontier for basis pursuit solutions. SIAM J. Sci. Comput. **31**(2), 890–912 (2008)
7. Bertsekas, D.: Constrained Optimization and Lagrange Multiplier Methods. Athena scientific series in optimization and neural computation, 1st edn. Athena Scientific, Massachusetts (1996)
8. Bertsekas, D.: Nonlinear Programming. Athena Scientific, Massachusetts (1999)
9. Bioucas-Dias, J.M.: Fast GEM wavelet-based image deconvolution algorithm. In: IEEE International Conference on Image Processing- ICIP 2003, vol. 2, pp. 961–964 (2003)
10. Bioucas-Dias, J.M.: Bayesian wavelet-based image deconvolution: a GEM algorithm exploiting a class of heavy-tailed priors. IEEE Trans. Image Process. **15**(4), 937–951 (2006)
11. Bioucas-Dias, J.M., Figueiredo, M.A.T.: A new TwIST: two-step iterative shrinkage/thresholding algorithms for image restoration. IEEE Trans. Image Process. **16**(12), 2992–3004 (2007)
12. Bioucas-Dias, J.M., Figueiredo, M.A.T.: Multiplicative noise removal using variable splitting and constrained optimization. IEEE Trans. Image Process. **19**(7), 1720–1730 (2010)
13. Boyd, S., Vandenberghe, L.: Convex Optimization. Cambridge University Press, USA (2004)
14. Bregman, L.: The relaxation method of finding the common point of convex sets and its application to the solution of problems in convex programming. USSR Comput. Math. Math. Phys. **7**(3), 200–217 (1967)
15. Bregman, L.M.: The method of successive projection for finding a common point of convex sets. Sov. Math. Dokl. **6**, 688–692 (1965)
16. Candes, E.J., Romberg, J.K.: Signal recovery from random projections. In: Proceedings of SPIE Computational Imaging III, vol. 5674, pp. 76–86. San Jose (2005)
17. Combettes, P.L.: Iterative construction of the resolvent of a sum of maximal monotone operators. J. Convex Anal. **16**, 727–748 (2009)
18. Combettes, P.L., Pesquet, J.C.: A proximal decomposition method for solving convex variational inverse problems. Inverse Probl. **24**(6), 1–27 (2008)
19. Combettes, P.L., Pesquet, J.C.: Proximal splitting methods in signal processing. In: FixedPoint Algorithms for Inverse Problems in Science and Engineering, pp. 185–212. Springer, New York (2011)
20. Combettes, P.L., Wajs, V.R.: Signal recovery by proximal forward-backward splitting. Multiscale Model. Simul. **4**(4), 1168–1200 (2005)
21. Daubechies, I., Defrise, M., De Mol, C.: An iterative thresholding algorithm for linear inverse problems with a sparsity constraint. Commun. Pure Appl. Math. **57**(11), 1413–1457 (2004)
22. Deka, B., Datta, S.: High throughput MR image reconstruction using compressed sensing. ICVGIP **14**, 89:1–89: 6 (2014). ACM, Bangalore, India
23. Demmel, J.W.: Applied Numerical Linear Algebra. Society for industrial and applied mathematics, USA (1997)
24. Dolui, S.: Variable splitting as a key to efficient cient image reconstruction. Ph.D. thesis, Electrical and Computer Engineering, University of Waterloo (2012)
25. Eckstein, J.: Splitting methods for monotone operators with applications to parallel optimization. Ph.D. thesis, Department of Civil Engineering, Massachusetts Institute of Technology (1989)
26. Eckstein, J., Bertsekas, D.P.: On the douglas-rachford splitting method and the proximal point algorithm for maximal monotone operators. Math. Program. **55**, 293–318 (1992)
27. Figueiredo, M., Bioucas-Dias, J., Nowak, R.: Majorization minimization algorithms for wavelet-based image restoration. IEEE Trans. Image Process. **16**(12), 2980–2991 (2007)
28. Figueiredo, M., Nowak, R.: An EM algorithm for wavelet-based image restoration. IEEE Trans. Image Process. **12**(8), 906–916 (2003)
29. Figueiredo, M., Nowak, R., Wright, S.: Gradient projection for sparse reconstruction: Application to compressed sensing and other inverse problems. IEEE J. Sel. Top. Signal Process. **1**(4), 586–597 (2008)

30. Gabay, D., Mercier, B.: A dual algorithm for the solution of nonlinear variational problems via finite element approximation. Comput. Math. Appl. **2**(1), 17–40 (1976)
31. Glowinski, R., Le Tallec, P.: Augmented Lagrangian and Operator-splitting Methods in Nonlinear Mechanics. SIAM studies in applied mathematics. Society for Industrial and Applied Mathematics, Philadelphia (1989)
32. Glowinski, R., Marrocco, A.: Sur lapproximation, par elements finis dordre un, et la resolution, parpenalisation-dualite, dune classe de problems de dirichlet non lineares. Revue Francaise dAutomatique, Informatique, et Recherche Op erationelle **9**, 41–76 (1975)
33. Goldstein, T., Osher, S.: The split bregman method for L_1-regularized problems. SIAM J. Imaging Sci. **2**(2), 323–343 (2009)
34. Grippo, L., Sciandrone, M.: Nonmonotone globalization techniques for the Barzilai-Borwein gradient method. Comput. Optim. Appl. **23**(2), 143–169 (2002)
35. Hale, E., Yin, W., Zhang, Y.: A fixed-point continuation method for L_1-regularized minimization with applications to compressed sensing. Technical report. Rice University, CAAM (2007)
36. Hestenes, M.: Multiplier and gradient methods. J. Optim. Theory Appl. **4**(5), 303–320 (1969)
37. Hestenes, M.R., Stiefel, E.: Methods of conjugate gradients for solving linear systems. J. Res. Natl. Bur. Stand. **49**, 409–436 (1952)
38. Huang, J., Zhang, S., Li, H., Metaxas, D.N.: Composite splitting algorithms for convex optimization. Comput. Vis. Image Underst. **115**(12), 1610–1622 (2011)
39. Huang, J., Zhang, S., Metaxas, D.N.: Efficient MR image reconstruction for compressed MR imaging. Med. Image Anal. **15**(5), 670–679 (2011)
40. Kim, S., Koh, K., Lustig, M., Boyd, S., Gorinevsky, D.: An interior-point method for largescale L_1-regularized least squares. IEEE J. Sel. Top. Signal Process. **1**(4), 606–617 (2008)
41. Lustig, M.: Sparse MRI. Ph.D. thesis, Electrical Engineering, Stanford University (2008)
42. Lustig, M., Donoho, D., Pauly, J.M.: Sparse MRI: the application of compressed sensing for rapid MR imaging. Magn. Reson. Med. **58**, 1182–1195 (2007)
43. Ma, S., Yin, W., Zhang, Y., Chakraborty, A.: An efficient algorithm for compressed MR imaging using total variation and wavelets. In: IEEE Conference on Computer Vision and Pattern Recognition (CVPR 2008), pp. 1–8. Anchorage, AK (2008)
44. Majumdar, A.: Compressed Sensing for Magnetic Resonance Image Reconstruction. Cambridge University Press, India (2015)
45. Mallat, S., Zhang, Z.: Matching pursuits with time-frequency dictionaries. IEEE Trans. Signal Process. **41**, 3397–3415 (1993)
46. Moreau, J.J.: Fonctions convexes duales et points proximaux dans un espace hilbertien. Comptes Rendus de lAcad emie des Sciences (Paris), S erie A **255**, 2897–2899 (1962)
47. Nesterov, Y.: A method of solving a convex programming problem with convergence rate O(1/sqr(k)). Sov. Math. Dokl. **27**, 372–376 (1983)
48. Osher, S., Burger, M., Goldfarb, D., Xu, J., Yin, W.: An iterative regularization method for total variation-based image restoration. Multiscale Model. Simul. **4**(2), 460–489 (2005)
49. Parikh, N., Boyd, S.: Proximal algorithms. Found. Trends Optim. **1**(3), 127–239 (2014)
50. Pati, Y.C., Rezaiifar, R., Pati, Y.C., Rezaiifar, R., Krishnaprasad, P.S.: Orthogonal Matching Pursuit: Recursive Function Approximation with Applications to Wavelet Decomposition. pp. 40–44(1993)
51. Powell, M.J.D.: A method for nonlinear constraints in minimization problems. In: Fletcher, R. (ed.) Optimization, pp. 283–298. Academic Press, New York (1969)
52. Rudin, L.I., Osher, S., Fatemi, E.: Nonlinear total variation based noise removal algorithms. Phys. D **60**, 259–268 (1992)
53. Wang, Y., Yang, J., Yin, W., Zhang, Y.: A new alternating minimization algorithm for total variation image reconstruction. SIAM J. Imaging Sci. **1**(3), 248–272 (2008)
54. Wright, S.J., Nowak, R.D., Figueiredo, M.A.T.: Sparse reconstruction by separable approximation. IEEE Trans. Signal Process. **57**(7), 2479–2493 (2009)
55. Xiao, Y., Yang, J., Yuan, X.: Alternating algorithms for total variation image reconstruction from random projections. Inverse Probl. Imaging (IPI) **6**(3), 547–563 (2012)

56. Yang, A.Y., Ganesh, A., Zhou, Z., Sastry, S., Ma, Y.: A review of fast L_1-minimization algorithms for robust face recognition. CoRR (2010). arXiv:1007.3753
57. Yang, J., Zhang, Y.: Alternating direction algorithms for L_1-problems in compressive sensing. SIAM J. Sci. Comput. **33**(1), 250–278 (2011)
58. Yang, J., Zhang, Y., Yin, W.: A fast alternating direction method for TVL_1-L_2 signal reconstruction from partial Fourier data. IEEE J. Sel. Top. Signal Process. **4**(2), 288–297 (2010)
59. Yin, W., Osher, S., Goldfarb, D., Darbon, J.: Bregman iterative algorithms for L_1-minimization with applications to compressed sensing. SIAM J. Imaging Sci., 143–168 (2008)
60. Youla, D., Webb, H.: Image restoration by the method of convex projections: Part 1-theory. IEEE Trans. Med. Imaging **1**(2), 81–94 (1982)

Chapter 4
Performance Evaluation of CS-MRI Reconstruction Algorithms

Abstract Performances of various compressed sensing reconstruction algorithms are compared under a common simulation environment with different real and synthetic MRI datasets. From experimental results, it has been observed that composite splitting based algorithms outperform others in terms of reconstruction quality, CPU time, and visual results. Additionally, to demonstrate the effectiveness of iterative reweighting an adaptive weighting scheme is combined with a fast composite splitting algorithm and its improvements are also presented.

4.1 Introduction

In CS-MRI, quality of reconstructed images mainly depends on how well a particular CS reconstruction model has been able to converge to the true solution with a given stopping criteria. Computational time is another major concern for CS-MRI reconstruction. For clinical applications one needs to reconstruct hundreds of images from multi-slice multichannel data that too within a couple of minutes without compromising the diagnostic information embedded in them. Due to these reasons, evaluation of different CS-MRI reconstruction techniques is vital in terms of their quality of reconstruction and computational time.

It is worth mentioning that for fair comparison an uniform simulation environment should be used. As discussed in the previous chapter algorithms have been divided into four broad categories and compared them within its own category. The evaluation process was carried out with both synthetic and real MRI datasets.

4.2 Simulation Setup

To evaluate performance we use a common experimental setup for all algorithms. All the experiments are performed in the MATLAB(R2012b) environment on a 3.40 GHz Intel $i7$ core CPU with 2 GB of RAM, 32 bit OS. We obtain MATLAB source codes

B. Deka and S. Datta, *Compressed Sensing Magnetic Resonance Image Reconstruction Algorithms*, Springer Series on Bio- and Neurosystems 9,
https://doi.org/10.1007/978-981-13-3597-6_4

of various algorithms, namely, the NCG [16] from,[1] the TNIPM [15] from,[2] the IST [7] from,[3] the TwIST [3] from,[4] the FISTA [2] from,[5] the SpaRSA [19] from,[6] the GPSR [10] from,[3] the SALSA [1] from,[7] the RecPF [20] from,[8] the TVCMRI [17] from,[9] the CSA [13] from[10] and the FCSA [14] from.[10]

4.2.1 MRI Database Selection

Performances of different CS-MRI reconstruction algorithms are tested on both synthetic as well as in vivo MR images. Two real MR image datasets are collected from the GNRC Hospital, Guwahati, India.[11] Figure 4.1a shows a 2D single slice Sag T2 TOP L.S. Spine MRI data, while Fig. 4.1b shows a randomly selected slice from 2D multi-slice 3D BRAVO T1 HR Brain MR images. First one was acquired using a GE 1.5T signa HDxt scanner with following parameters: TR/TE: 3520/111.044 ms, slice thickness: 4 mm, spacing between scans: 5 mm, sampling (%): 100, and Flip angle: 90°. On the other hand, the brain MRI was acquired using a GE 1.5T signa HDxt scanner with the following parameters: TR/TE: 14.912/6.456 ms, slice thickness: 1.2 mm, spacing between scans: 1.2 mm, sampling (%): 100, and Flip angle: 15°.

Some in vivo MR images are also taken from a publicly available database at.[12] A set of 3D fast spin echo (FSE) Knee MR images are collected from this dataset and an example is shown in Fig. 4.1c. They were collected from a GE HDx 3T scanner with following parameters, field of view (FOV): 160 × 160 mm, TR/TE: 1550/25.661 ms, slice thickness (ST): 0.6 mm, spacing between scans: 0 mm and matrix size: 320 × 320. Also, several brain MR images are taken from *MRI of Trinidad and Tobago Limited*.[13] One such Brain MR image is shown in Fig. 4.1d. These representative images are used as *test images* for demonstration of various experimental results.

To evaluate performances of various algorithms on synthetic MR images and also to compare them with those obtained from in vivo images, we generate some brain MR images using the BrainWeb simulator.[14] Synthetic data are commonly

---

[1] http://www.eecs.berkeley.edu/~mlustig/software/sparseMRI_v0.2.tar.gz.

[2] http://www.stanford.edu/~boyd/l1_1s/.

[3] http://www.lx.it.pt/~mtf/GPSR/.

[4] http://www.lx.it.pt/~bioucas/TwIST/TwIST.htm.

[5] http://www.eecs.berkeley.edu/~yang/software/l1benchmark/l1benchmark.zip.

[6] http://www.lx.it.pt/~mtf/SpaRSA/.

[7] http://cascais.lx.it.pt/~mafonso/salsa.html.

[8] http://www.caam.rice.edu/~optimization/L1/RecPF/.

[9] http://www1.se.cuhk.edu.hk/~sqma/codes/TVCMRI_pub.zip.

[10] http://ranger.uta.edu/~huang/codes/FCSA_MRI1.0.rar.

[11] http://www.gnrchospitals.com.

[12] http://mridata.org/fullysampled/knees.

[13] http://mritnt.com/education-centre/common-uses/mri-of-the-brain/.

[14] http://brainweb.bic.mni.mcgill.ca/brainweb.

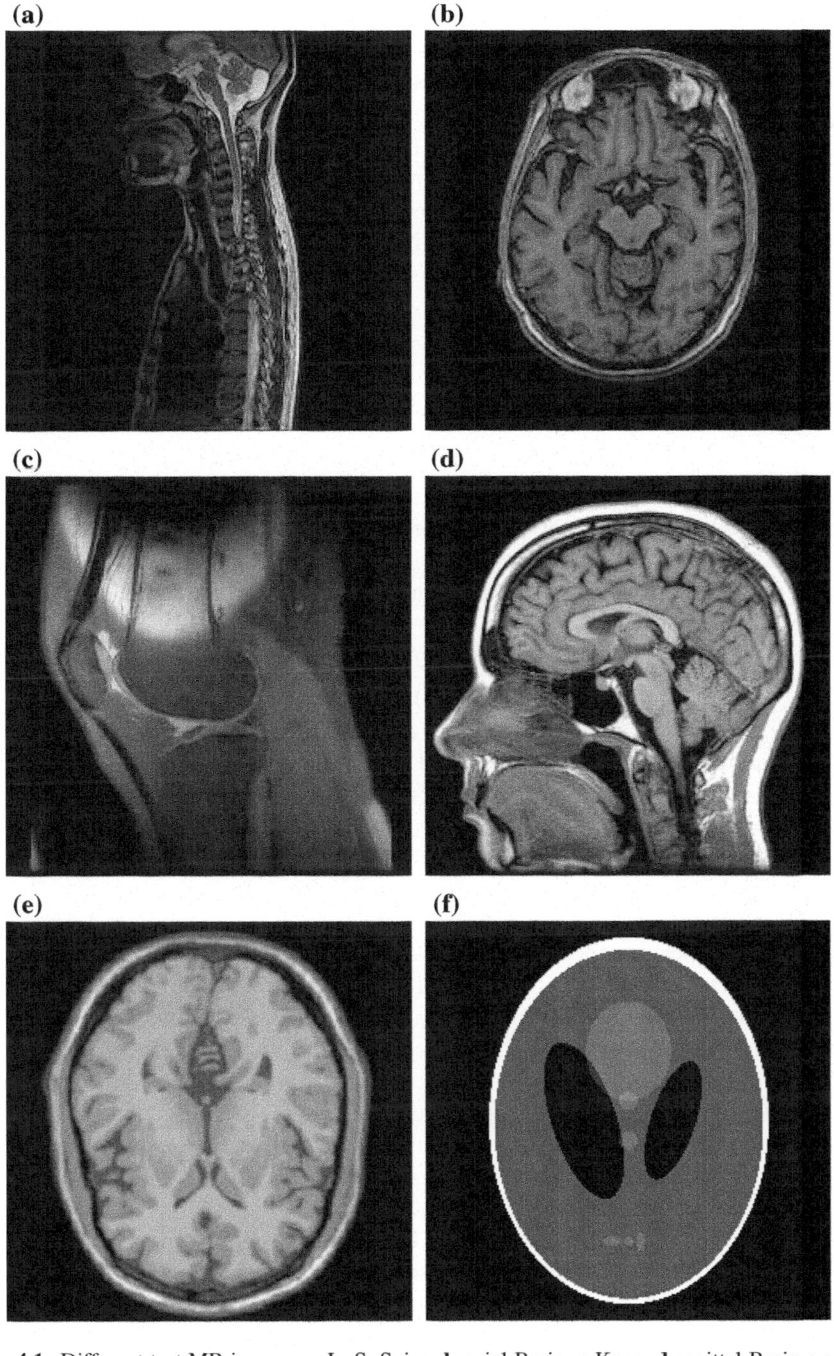

Fig. 4.1 Different test MR images: **a** L. S. Spine, **b** axial Brain, **c** Knee, **d** sagittal Brain **e** axial Brain using BrainWeb simulator and **f** Phantom

used to evaluate performances of different algorithms in the noise-free case, i.e., where ground truth is known. On the other hand, in vivo MR images are used to test performances of various algorithms in practical applications where data itself contain variable amounts of noise and images are of low contrast compared to synthetic images.

We have simulated 2D multi-slice brain MR images with following parameters, pulse sequence: SFLASH, TR/TE: 18/10 ms, Flip angle: 30°, Image Type: magnitude, Noise level: 0%, INU field: field A and INU level: 20%. One synthetic brain MR image is shown in Fig. 4.1e. Finally, we also use the MATLAB phantom image shown in Fig. 4.1f for our simulation.

4.2.2 Selection of Parameters

The performance of an algorithm depends on multiple factors, including stopping criteria, selection of parameters, and algorithm implementation. Depending on the algorithm, an appropriate stopping criterion is essential for its convergence to a local or global optimum value. For fair comparisons, we use the following common stopping criterion, i.e.,

$$\frac{\left\| f\left(\mathbf{x}^{(k+1)}\right) - f\left(\mathbf{x}^{(k)}\right)\right\|_2}{\left\| f\left(\mathbf{x}^{(k)}\right)\right\|_2} \leqslant \varepsilon \tag{4.1}$$

where tolerance ε is taken 10^{-4} in the MATLAB simulation. Besides stopping criteria, convergence of an algorithm highly depends on the step size of the iteratively updating parameters. If the step size of the updating parameter is not properly chosen then the algorithm may terminate before reaching near the optimum solution or it takes longer times for convergence.

Reconstruction performances of different algorithms are evaluated by some widely accepted performance indices, namely, the peak signal-to-noise ratio (PSNR), the mean structural similarity index (MSSIM) [18] and the CPU time. PSNR is used to evaluate the quality of reconstructed images by measuring the mean squared error (MSE) which indicates the deviation of reconstructed images from the ground truth. Although it is a very good index when the ground truth is known but it sometimes fails to show the actual difference between two images if the ground truth is not known. It is observed that two images may have the same MSE value but visually they look very much dissimilar. Human visual system is more sensitive to the gross structural information present in the image. MSSIM compares structural similarities between two images instead of mere pixel differences. It returns a 0 which indicates no similarity, and 1 which indicates exact similarity between the two images. In case of CS-MRI, it is used to evaluate detailed quality of reconstructed MR images from highly undersampled Fourier data. On the other hand, the CPU time is used to evaluate computational costs of various algorithms. However, it is a machine-dependent parameter, it may vary depending on the machine configurations. But, it definitely gives relative information about computational costs of different algorithms.

4.3 Performance Evaluation

Evaluation of various CS-MRI reconstruction algorithms is done using six test images including four real and two simulated MR images as described above. Algorithms have been classified into four categories based on their problem-solving approaches. Since the purpose here is to compare reconstruction performances only, we consider a common random undersampling scheme based on variable density undersampling pattern as shown in Fig. 1.8b for all algorithms.

Performances in terms of PSNR (in dB) on the axial Brain image are shown in Fig. 4.2. Separate plots for different categories of algorithms, namely, the *operator splitting*, the *variable splitting*, the *composite splitting*, and the *non-splitting* based algorithms are shown for better understanding. It is clearly observed that the TVCMRI, the RecPF, the ALM-CSD, and the NCG give the best reconstruction performance in their respective categories in terms of the PSNR. Further, Figs. 4.3, 4.4, 4.5, 4.6, and 4.7 show results for the L.S. Spine, the Knee, the sagittal Brain, the BrainWeb, and the Phantom images, respectively. Similar observations are also noted for other test images as observed for the axial Brain. On the other hand, out of different categories, the composite splitting based algorithms perform better compared to other categories because it exploits advantages of both the operator and variable splitting techniques.

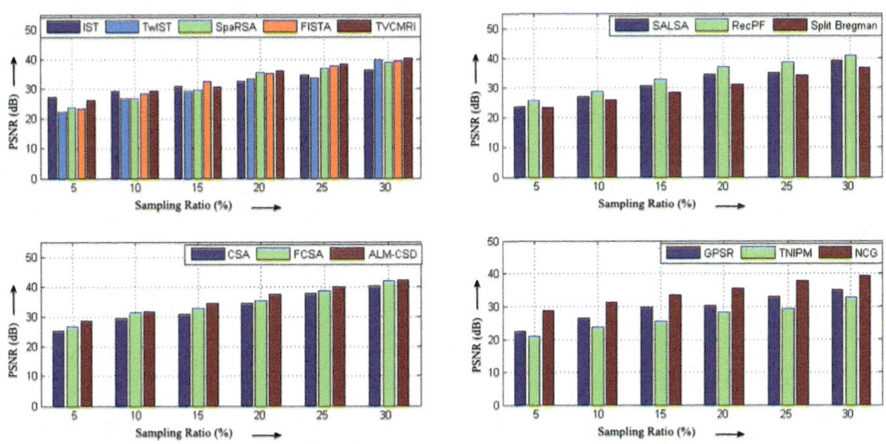

Fig. 4.2 Comparison of PSNR (in dB) for the reconstruction of the axial Brain image using various algorithms with varying undersampling ratio. First row left to right: results of operator splitting and variable splitting based algorithms. Second row left to right: results of composite splitting and non-splitting based algorithms

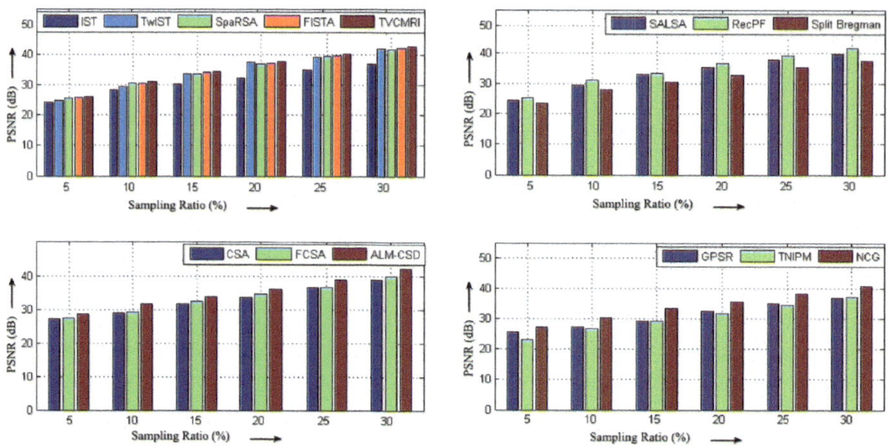

Fig. 4.3 Comparison of PSNR (in dB) for the reconstruction of the L. S. Spine image using various algorithms with varying undersampling ratio. First row left to right: results of operator splitting and variable splitting based algorithms. Second row left to right: results of composite splitting and non-splitting based algorithms

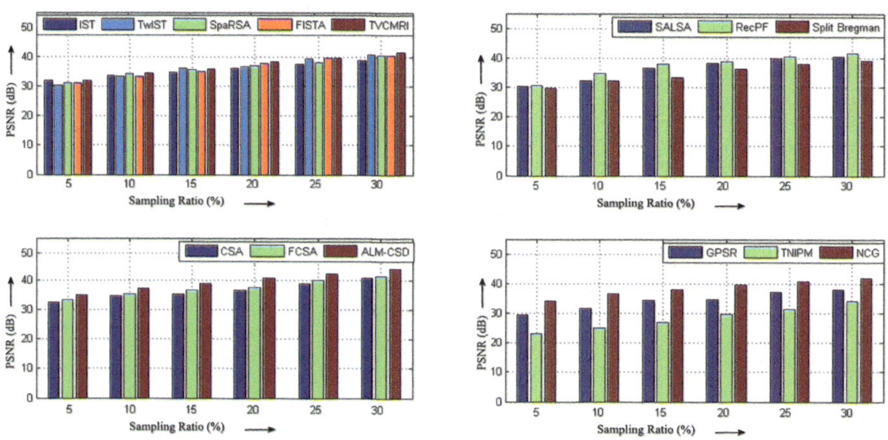

Fig. 4.4 Comparison of PSNR (in dB) for the reconstruction of the Knee image using various algorithms with varying undersampling ratio. First row left to right: results of operator splitting and variable splitting based algorithms. Second row left to right: results of composite splitting and non-splitting based algorithms

Performances of various algorithms in terms of MSSIM on the sagittal Brain image are shown in Fig. 4.8. It also shows similar results as observed in terms of the PSNR. Again, among the four different categories, the composite splitting based algorithms give better MSSIM values as compared to others. Similarly, results are also obtained for the simulated BrainWeb image and shown in Fig. 4.9 which are in line with those for real images. Now, from these observations based on PSNR and MSSIM, it may

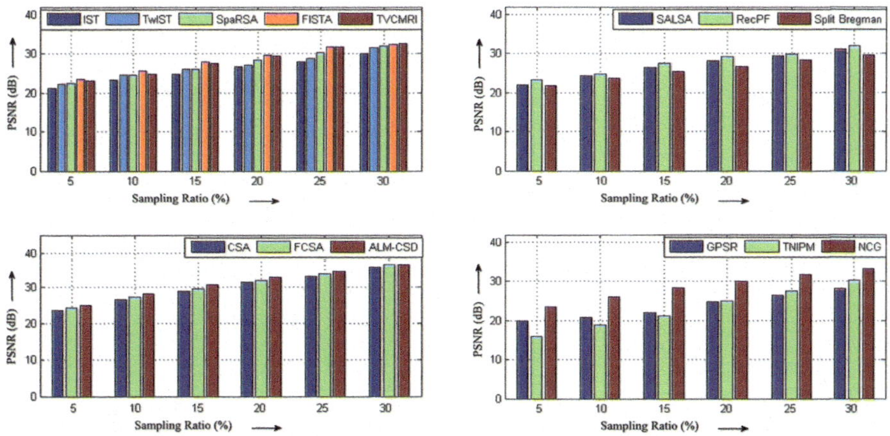

Fig. 4.5 Comparison of PSNR (in dB) for the reconstruction of the sagittal Brain image using various algorithms with varying undersampling ratio. First row left to right: results of operator splitting and variable splitting based algorithms. Second row left to right: results of composite splitting and non-splitting based algorithms

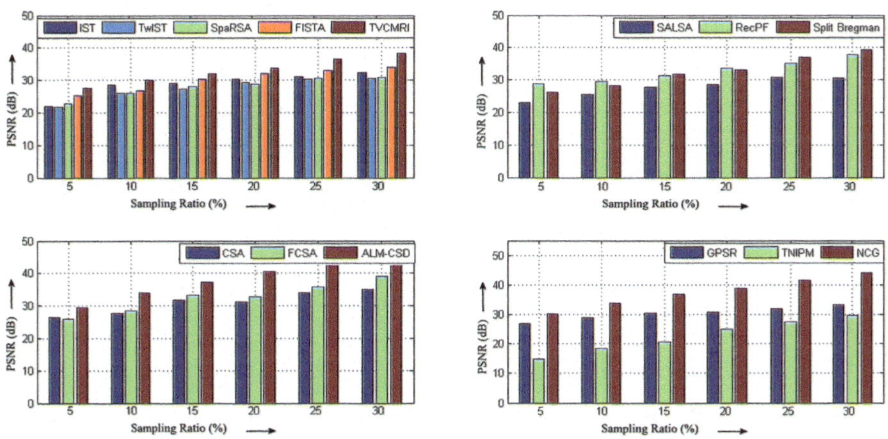

Fig. 4.6 Comparison of PSNR (in dB) for the reconstruction of the BrainWeb image using various algorithms with varying undersampling ratio. First row left to right: results of operator splitting and variable splitting based algorithms. Second row left to right: results of composite splitting and non-splitting based algorithms

be concluded that the composite splitting category gives better quality of CS-MRI reconstruction than their counterparts in other remaining categories.

In addition to the quality of reconstructions, computational costs of various algorithms for convergence are evaluated in terms of CPU time. Results for various algorithms of different categories using the sagittal Brain, the BrainWeb, and the Phantom images are shown in Figs. 4.10, 4.11, and 4.12. The TVCMRI in the oper-

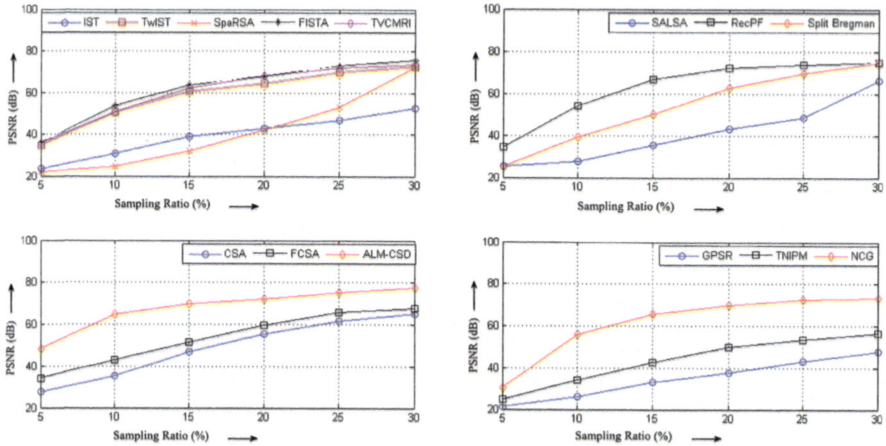

Fig. 4.7 Comparison of PSNR (in dB) for the reconstruction of the Phantom image using various algorithms with varying undersampling ratio. First row left to right: results of operator splitting and variable splitting based algorithms. Second row left to right: results of composite splitting and non-splitting based algorithms

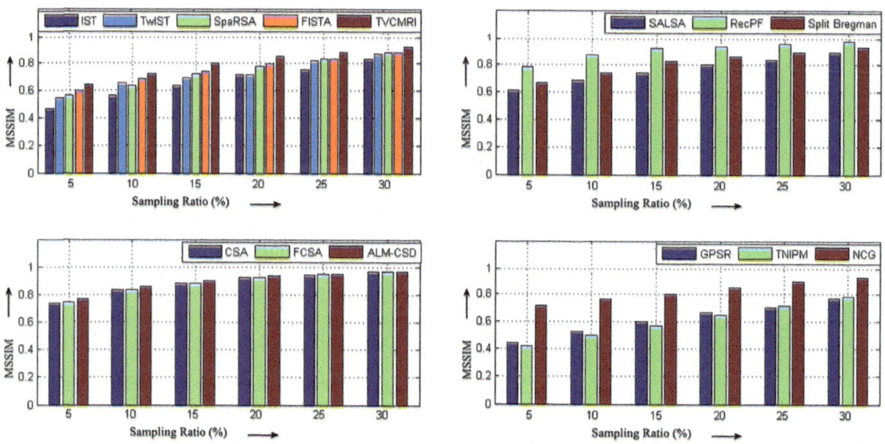

Fig. 4.8 Comparison of MSSIM for the reconstruction of the sagittal Brain image using various algorithms with varying undersampling ratio. First row left to right: results of operator splitting and variable splitting based algorithms. Second row left to right: results of composite splitting and non-splitting based algorithms

ator splitting category required the least CPU Time as compared to other algorithms for both the sagittal Brain and the Phantom images. But, in case of the BrainWeb image the FISTA and the TVCMRI required the similar CPU Time. In the variable splitting category, the RecPF required the least CPU Time as compared to other algorithms for different MR images. In case of composite splitting, the FCSA and the ALM-CSD require similar CPU Times for convergence for all three images.

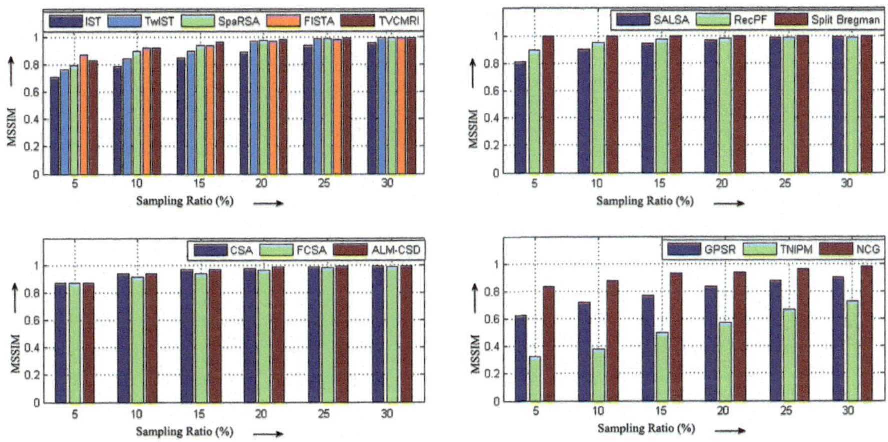

Fig. 4.9 Comparison of MSSIM for the reconstruction of the BrainWeb image using various algorithms with varying undersampling ratio. First row left to right: results of operator splitting and variable splitting based algorithms. Second row left to right: results of composite splitting and non-splitting based algorithms

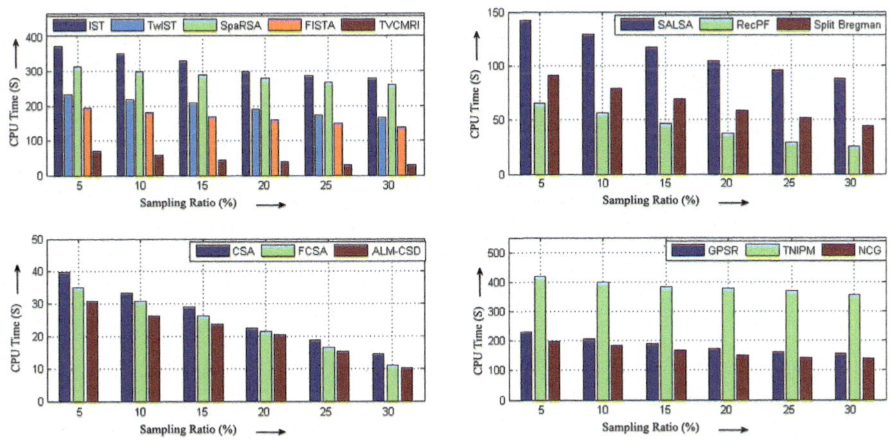

Fig. 4.10 Comparison of CPU time (in seconds) for the reconstruction of the sagittal Brain image using various algorithms with varying undersampling ratio. First row left to right: results of operator splitting and variable splitting based algorithms. Second row left to right: results of composite splitting and non-splitting based algorithms

Finally, for non-splitting category, the NCG algorithm required the least CPU Time compared to other algorithms. For the sagittal Brain and the Phantom images, the GPSR require very close CPU Time with that of the NCG. To summarize, it may be concluded that among all algorithms across different categories the FCSA and the ALM-CSD required the least CPU Time at different undersampling ratios across different test images.

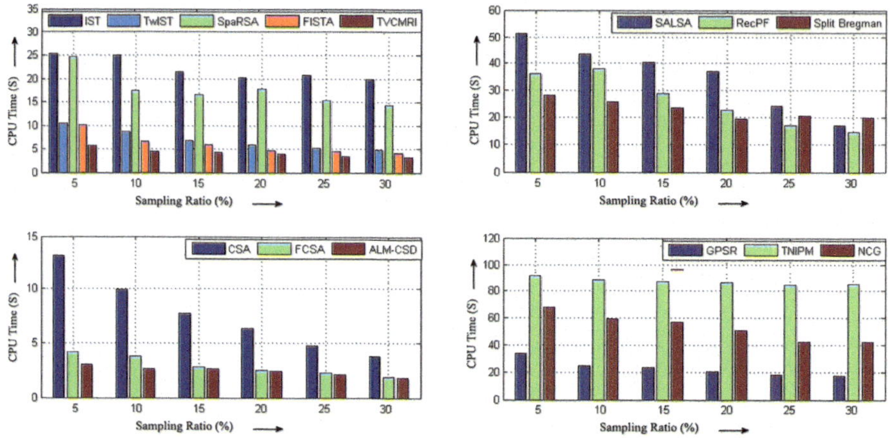

Fig. 4.11 Comparison of CPU time (in seconds) for the reconstruction of the BrainWeb image using various algorithms with varying undersampling ratio. First row left to right: results of operator splitting and variable splitting based algorithms. Second row left to right: results of composite splitting and non-splitting based algorithms

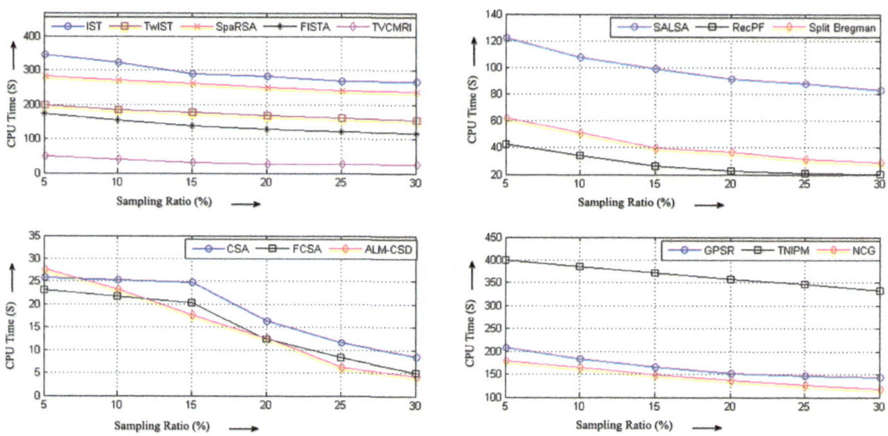

Fig. 4.12 Comparison of CPU time (in seconds) for the reconstruction of the Phantom image using various algorithms with varying undersampling ratio. First row left to right: results of operator splitting and variable splitting based algorithms. Second row left to right: results of composite splitting and non-splitting based algorithms

For better visualization of reconstruction results using various algorithms a cropped portion is zoomed in and shown in Figs. 4.13, 4.14, 4.15, 4.16, 4.17, 4.18, 4.19, and 4.20. It is observed that among different categories, composite splitting based algorithms yield better visual CS-MRI results with the least computational cost. Moreover, it is also researcher's friendly as one can easily modify any compos-

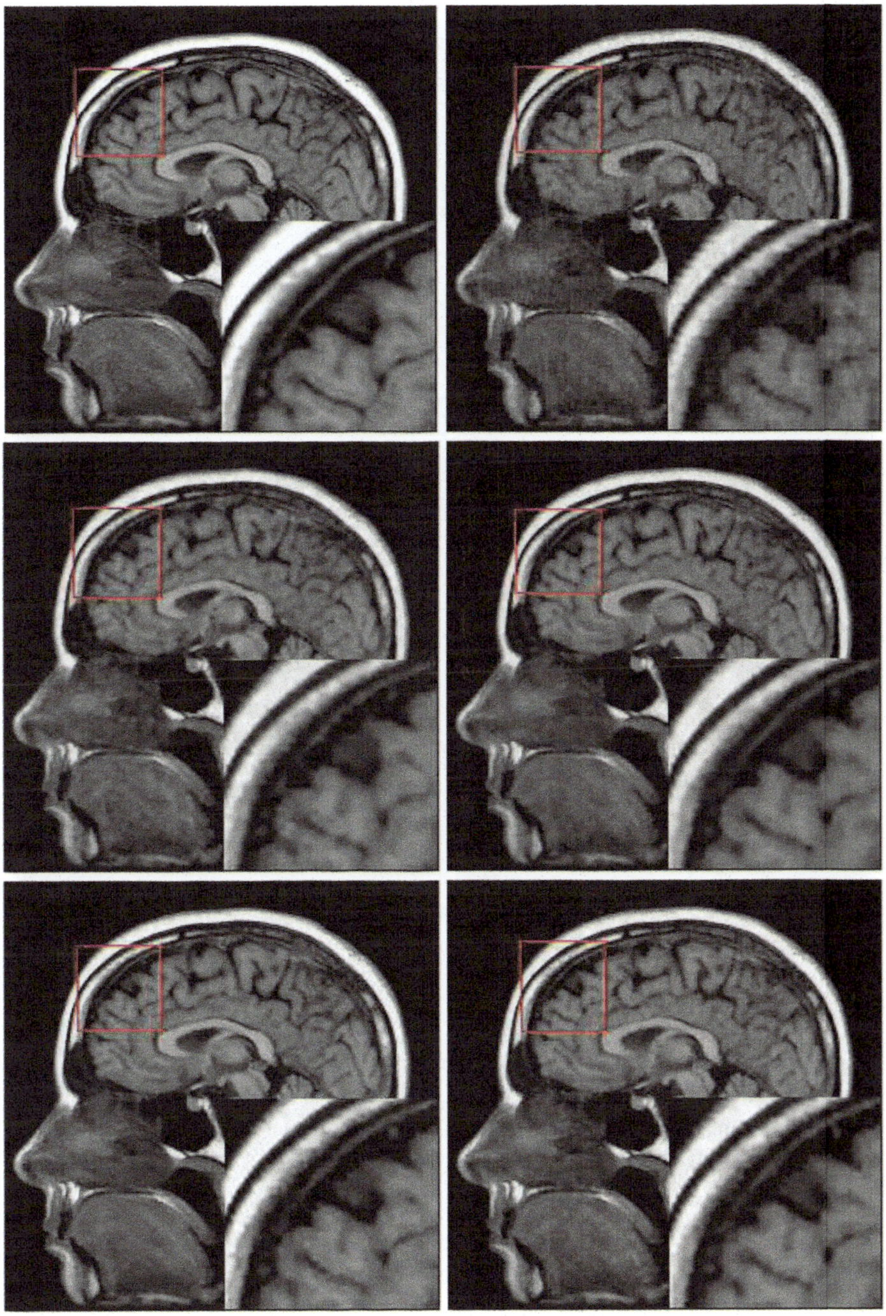

Fig. 4.13 Comparison of reconstructed Brain image using various operator splitting algorithms. Left to right and top to bottom: Original Brain image, reconstructed images using the IST, the TwIST, the TVCMRI, the FISTA, and the SpaRSA, respectively

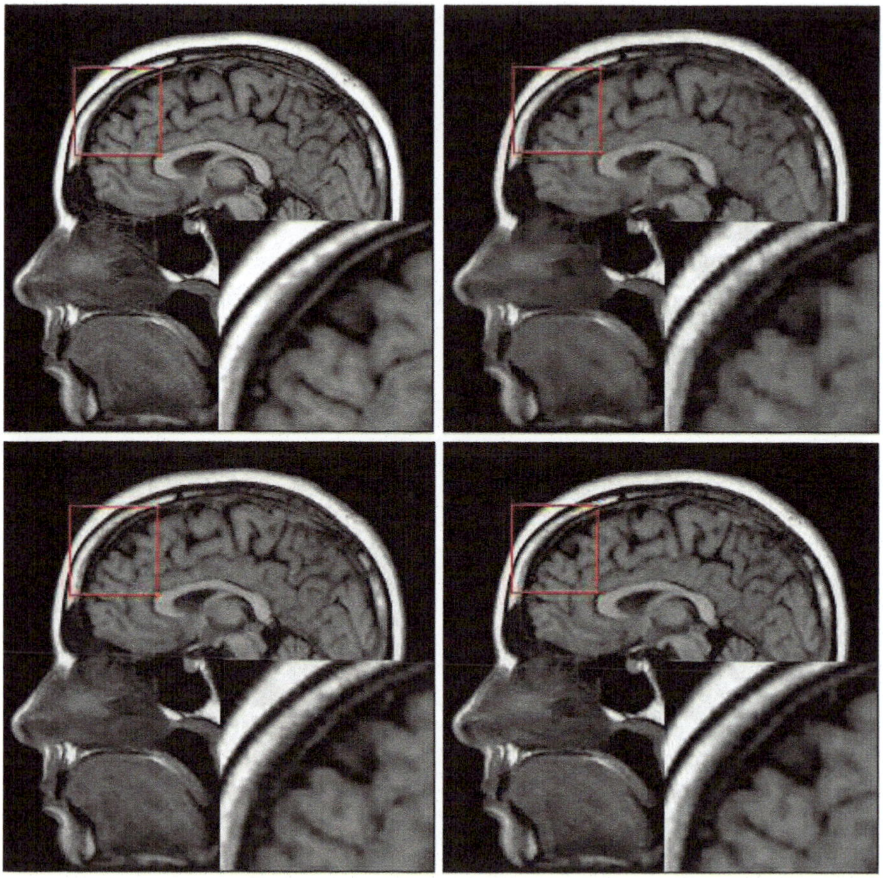

Fig. 4.14 Comparison of reconstructed Brain image using various variable splitting algorithms. Left to right and top to bottom: Original Brain image, reconstructed images using the split Bregman algorithm, the SALSA, and the RecPF, respectively

ite splitting model by adding additional sparsity promoting regularization terms for further improvement of the reconstruction quality.

The reader may refer to Table 4.1 as a summary of category-wise performance comparisons of different CS-MRI reconstruction approaches.

4.4 Experiments on Convergence

To test the convergence of algorithms from four different categories, first, two algorithms are randomly selected; one from the ℓ_1-ℓ_2 model (the IST) and other from the TV-ℓ_1-ℓ_2 model (the NCG). Then these two algorithms are run until convergence and

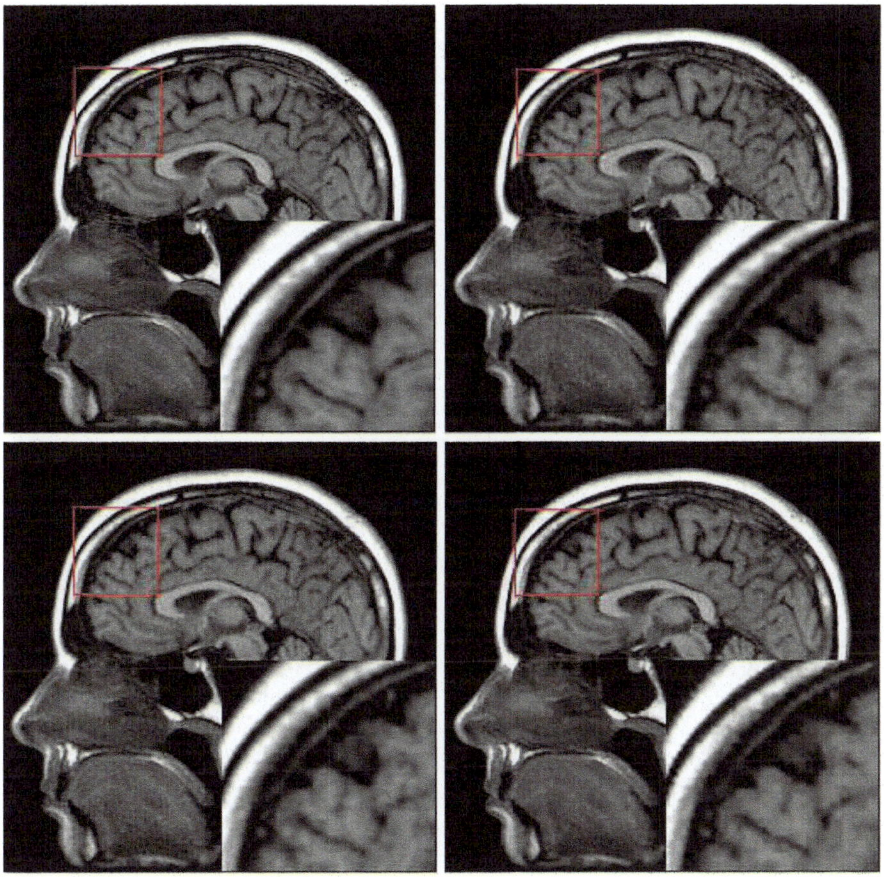

Fig. 4.15 Comparison of reconstructed Brain image using various composite splitting algorithms. Left to right and top to bottom: Original Brain image, reconstructed images using the CSA, the FCSA, and the ALM-CSD, respectively

corresponding objective function values are obtained which are to be used as target values for remaining algorithms belonging to the above two models. For an example, in the operator splitting category, the IST and the TVCMRI required approximately 40 s and 19 s, respectively, to reach their respective objective values. However, except them, other operator splitting algorithms reach the targeted objective function value in 2–3 s. Since the results are spread over a wide range, for better visualization of the evolution of objective functions of all algorithms together a small part of x-axis starting from origin is only consider. Similarly, objective function evolution for algorithms of other three categories may also be carried out. Consolidated results for all the categories are shown Fig. 4.21.

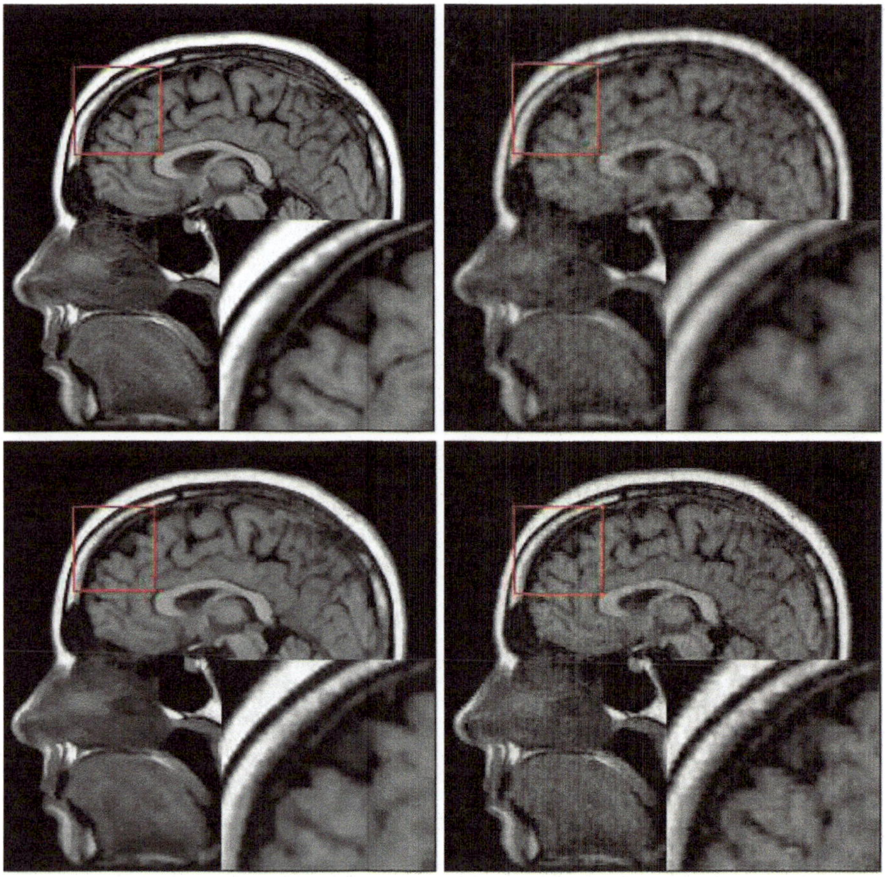

Fig. 4.16 Comparison of reconstructed Brain image using various non-splitting algorithms. Left to right and top to bottom: Original Brain image, reconstructed images using the TNIPM, the NCG and the GPSR, respectively

4.5 Performance Evaluation of Iteratively Weighted Algorithms

Iteratively reweighted least squares (IRLS) minimization is an alternative approach to solve ℓ_1-minimization problem [5, 8, 11, 12]. It targets to solve a ℓ_1-minimization problem by transforming it into a weighted ℓ_2 problem. In [11], IRLS was introduced as an approach to develop algorithms for solving sparse signal reconstruction problem by assuming that there is an initial solution which is close to the actual solution. In [8], theoretical justifications are given to prove that if the original ℓ_1-minimization problem has a unique, non-degenerate solution, then the IRLS procedure converges with first-order rate, i.e., linearly (exponentially) provided iterations

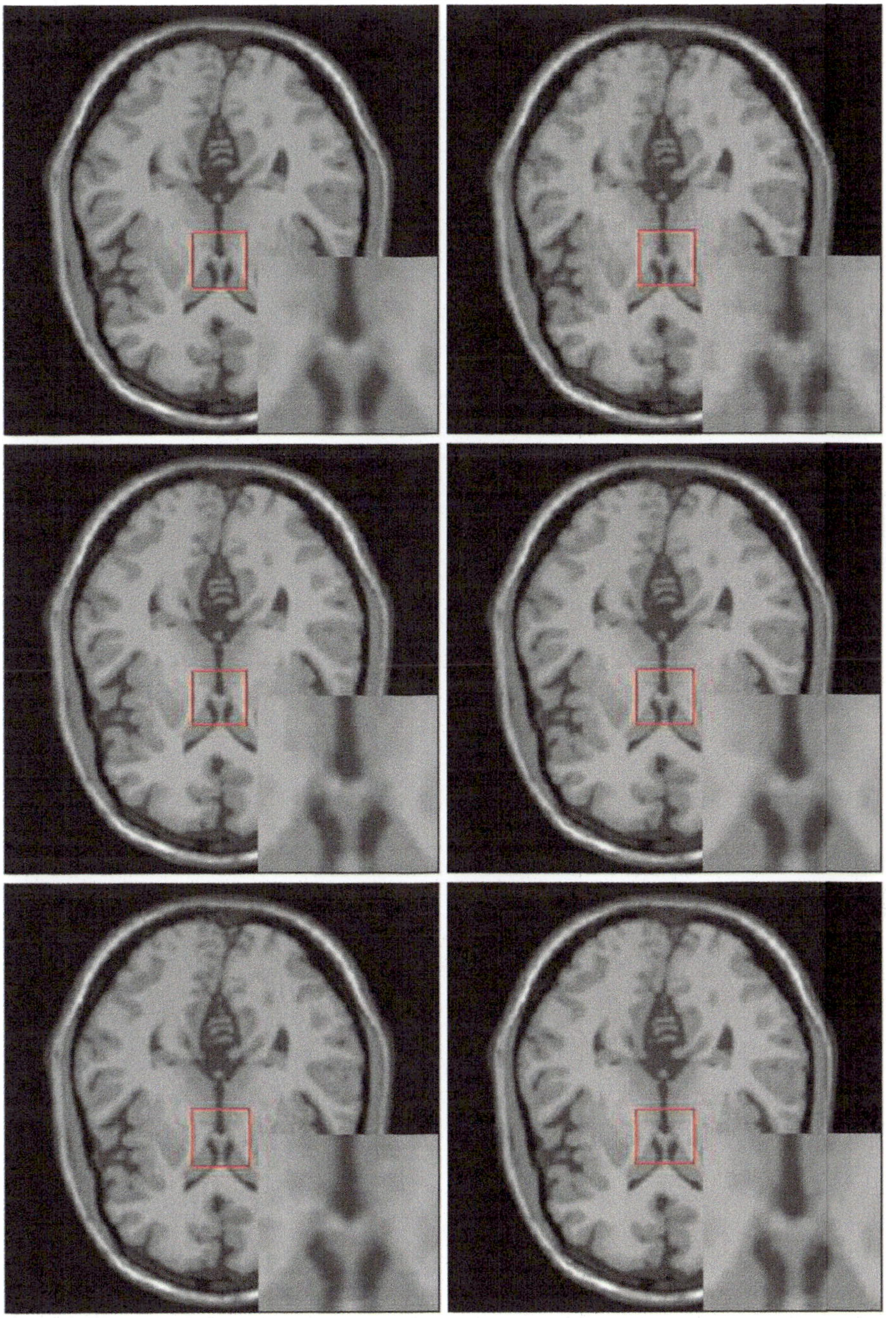

Fig. 4.17 Comparison of reconstructed BrainWeb image using various operator splitting algorithms. Left to right and top to bottom: Original BrainWeb image, reconstructed images using the IST, the TwIST, the TVCMRI, the FISTA, and the SpaRSA, respectively

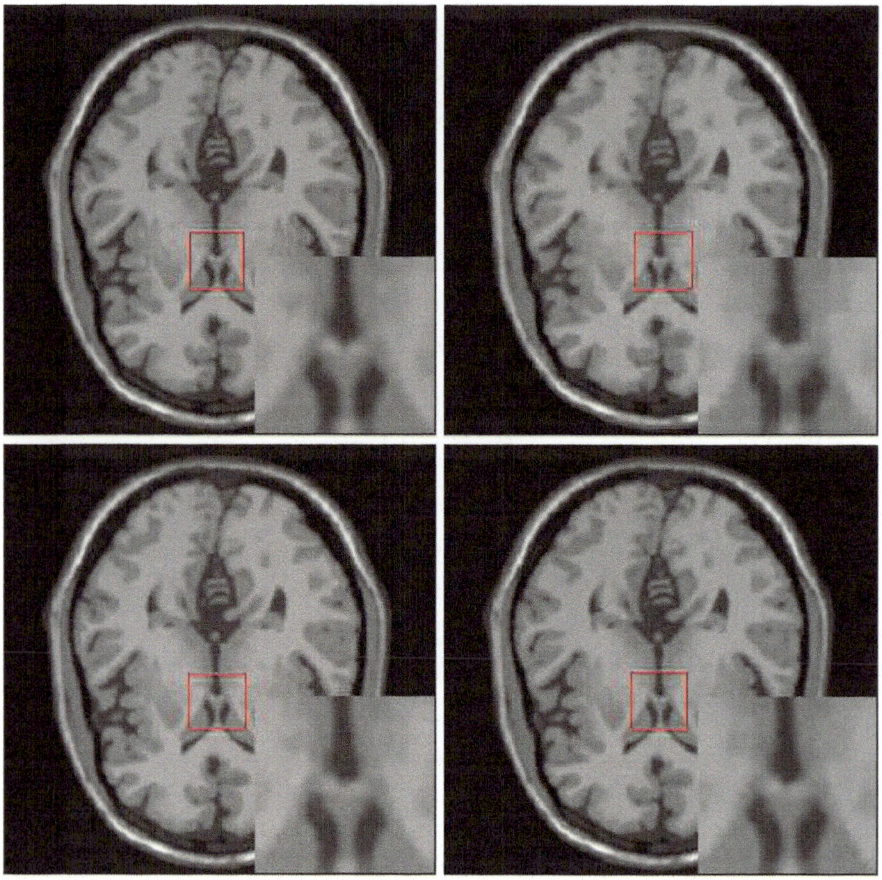

Fig. 4.18 Comparison of reconstructed BrainWeb image using various variable splitting algorithms. Left to right and top to bottom: Original BrainWeb image, reconstructed images using the split Bregman algorithm, the SALSA, and the RecPF, respectively

are begun with a vector sufficiently close to the solution. However, care must be taken in updating the weight vector from the solution of the current weighted least squares problem [8].

In [5], it has been shown that reweighting with the ℓ_1-norm minimization algorithms works as a catalyst; accelerate their speed of convergence with better solution that too at a reduced level of measurements than that of the ordinary ℓ_1 minimization. In case of ℓ_1-norm minimization for sparse representation, larger coefficients are relatively heavily penalized than smaller ones unlike the ℓ_0-norm minimization which is independent of magnitude of coefficients. To solve this imbalance, Candes et al. [5] proposed a new formulation known as the weighted ℓ_1-norm minimization to penalize the nonzero coefficients uniformly. The weighted ℓ_1 minimization problem can be defined as

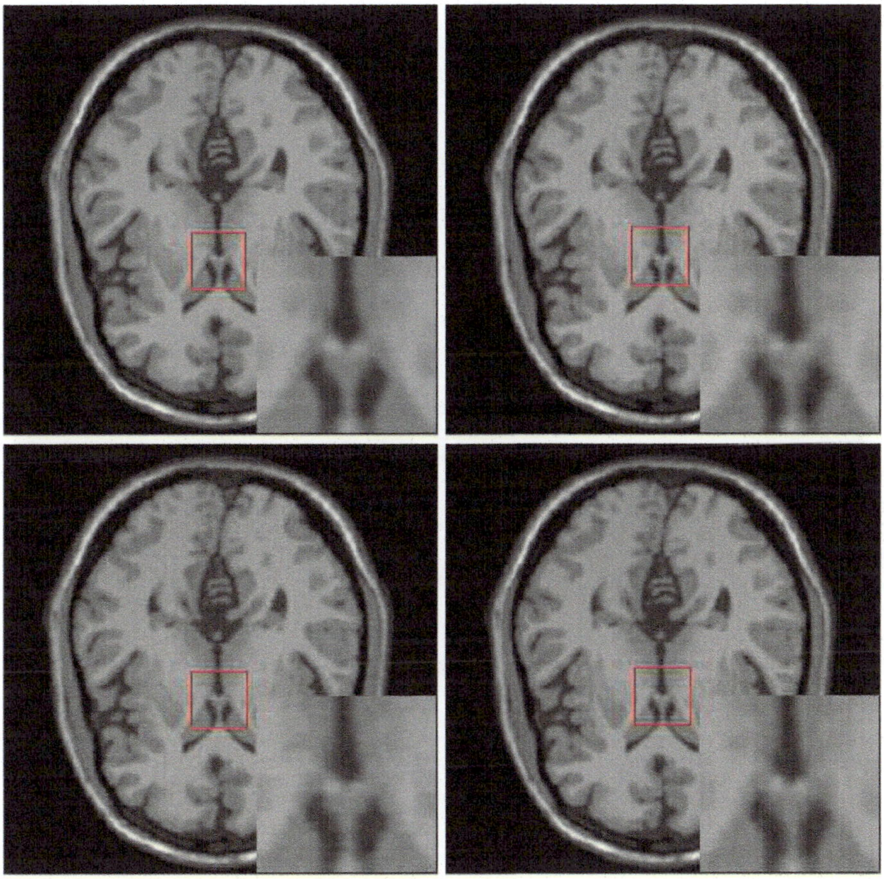

Fig. 4.19 Comparison of reconstructed BrainWeb image using various composite splitting algorithms. Left to right and top to bottom: Original BrainWeb image, reconstructed images using the CSA, the FCSA, and the ALM-CSD, respectively

$$\begin{aligned} \min_{\mathbf{x}} \quad & \sum_{i=1}^{n} w_i \, |x_i| \\ \text{subject to} \quad & \mathbf{y} = \mathbf{A}\mathbf{x} \end{aligned} \tag{4.2}$$

where $w_1, w_2, w_3, \ldots, w_n$ are positive weights. Ideally, these weights should minimize the magnitude dependency of the ℓ_1 minimization problem. Therefore, the weights are calculated in such a way that they are inversely proportional to the magnitude of the signal. Thus, weighted ℓ_1 minimization can be considered as a relaxed version of the ℓ_0 minimization.

To estimate a good weighting vector, at first, the weighted ℓ_1 minimization problem in Eq. 4.2 is solved with weighting vector $\mathbf{W}$ as an identity matrix which is

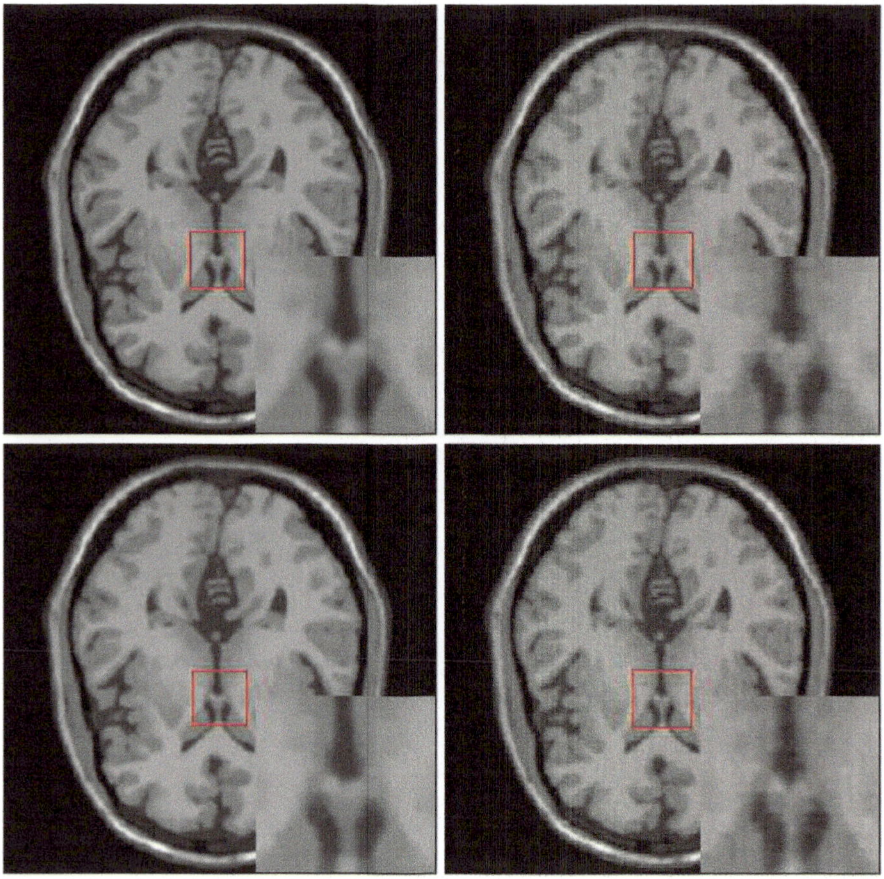

Fig. 4.20 Comparison of reconstructed BrainWeb image using various non-splitting algorithms. Left to right and top to bottom: Original BrainWeb image, reconstructed images using the TNIPM, the NCG and the GPSR, respectively

Table 4.1 Comparison of CS-MRI reconstruction performances

Evaluation criteria	CS-MRI reconstruction approach			
	Operator splitting	Variable splitting	Composite splitting	Non-splitting
PSNR	Moderate	Moderate	Highest	Lowest
MSSIM	Moderate	Moderate	Highest	Lowest
CPU time	Intermediate	Intermediate	Fast	Slow
Visual analysis	Good	Good	Best	Poor

equivalent to solve a conventional ℓ_1 minimization problem. Now, from this solution, a weighting vector is calculated which is used to solve the weighted ℓ_1 minimization problem which is called reweighted ℓ_1 minimization. On the other hand, in case of

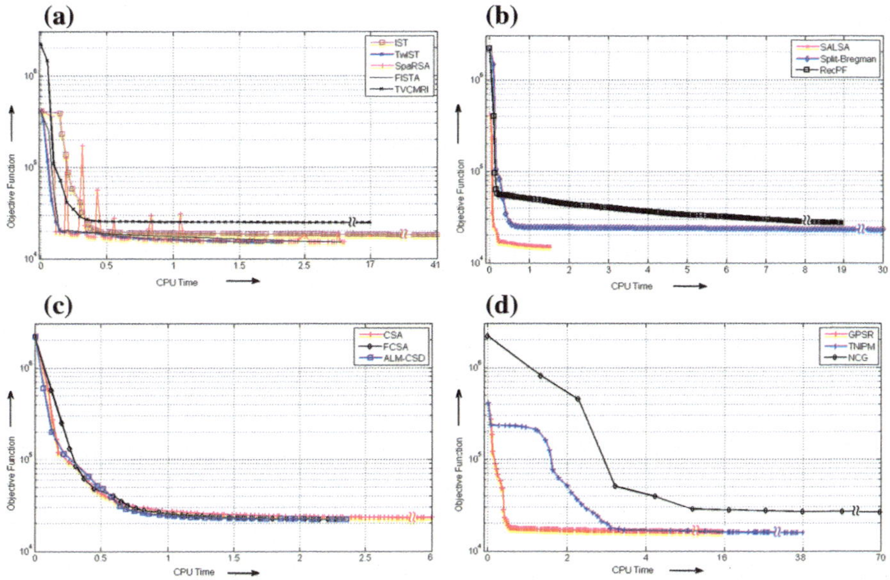

Fig. 4.21 Evolution of objective function with respect to CPU time of different categories of algorithms. **a** Operator splitting algorithms, **b** Variable splitting algorithms, **c** Composite splitting algorithms, and **d** Non-splitting algorithms

iteratively weighted ℓ_1 minimization the weighting vector is also changed within each iteration instead of computing the weighting vector first, and then solving the weighted ℓ_1 minimization problem with the already computed weighting vector from the previous solution.

The algorithmic steps of the iteratively weighted ℓ_1-norm minimization are shown in Algorithm 20. In this algorithm, $\mathbf{W}$ is a diagonal matrix and $w_i, \ldots, w_n$ are diagonal elements; the term $\varepsilon > 0$ is introduced to provide stability for zero-valued x_i. In Algorithm 20, step: 3 can be solved using any of the ℓ_1-norm minimization algorithms. Computational complexity for the weight calculation step is of the order of signal dimension, i.e., $O(n)$. Thus, per iteration the additional computational complexity for the iterative weighting scheme is only $O(n)$ which is same as the reweighted algorithm [5]. Although per iteration computational cost of the weighted ℓ_1 minimization problem is more compared to the standard ℓ_1 minimization, but it provides better reconstruction results with less number of measurements.

For example, ℓ_0 reconstruction of a k-sparse signal of length n would require not less than $2k$ measurements whereas the ℓ_1 reconstruction it would be of the order of $O\,(k\log(n))$ with a measurement matrix formed by sampling i.i.d. entries from the normal distribution [4, 5]. However, in the reweighted ℓ_1 reconstruction, the number of measurements required are higher than the ℓ_0 but less than the ℓ_1 reconstruction. Therefore, we may conclude that proper design of a weighting mechanism, the ℓ_1 reconstruction would be able to approach the ℓ_0 reconstruction in terms of number of measurements, which is one of the significant research findings for the CS-MRI.

To demonstrate this with experiments, we first assume a 1D sparse signal of length $n = 256$ with k nonzero elements generated randomly using unit-variance Gaussian distribution. The signal is acquired with a $m \times n$ measurement matrix which is generated using i.i.d Gaussian entries. To compare performances of different reweighted ℓ_1-minimization algorithms over the standard ℓ_1-minimization, we perform two sets of experiments. One is varying measurements m by keeping the sparsity level fixed and the other is varying the sparsity level k while keeping the number of measurements fixed. Figure 4.22 shows comparison of reconstruction performances of different reweighted ℓ_1 minimization algorithms with the standard ℓ_1-minimization. We run all algorithms on 100 randomly generated signals, each of size m and sparsity level k. We vary the number of measurements from 80 to 180 and kept k fixed at 50 for conducting the first experiment. Similarly, keeping the number of measurements fixed at 120, we vary k from 20 to 85 for the second experiment. We assume that the signal is perfectly reconstructed if $||\mathbf{x}_{\text{org}} - \mathbf{x}_{\text{rec}}||_{\ell_\infty} \leq 10^{-3}$. From Fig. 4.22a, we observe that the reweighted ℓ_1 algorithm gives a significant improvement over the ℓ_1-minimization alone. For perfect signal recovery, number of measurements m is reduced from 155 to 135 approximately. Moreover, for the iteratively weighted ℓ_1-minimization, measurements further reduce approximately to 125. Similarly, from Fig. 4.22b, we observe that there are significant reductions in the oversampling factors m/k for both reweighted and iteratively weighted ℓ_1 algorithms for perfect reconstruction compared to the ℓ_1. It is approximately $120/30 = 4$ for the ℓ_1 whereas for the reweighted and the iteratively weighted ℓ_1 algorithms they are approximately $120/40 = 3$ and $120/46 = 2.6$, respectively.

Algorithm 20 Iteratively weighted ℓ_1-minimization algorithm

Input: $\mathbf{A}$, $\mathbf{y}$, λ_1

1: $w_i^{(0)} \leftarrow \mathbf{1}, \quad i = 1....n$
2: **while** not converge **do**
3: $\quad \mathbf{x}^{(k)} \leftarrow \text{argmin} \left\|\mathbf{W}^{(k)}\mathbf{x}\right\|_1 \;$ subject to $\mathbf{y} = \mathbf{A}\mathbf{x}$
4: $\quad w_i^{(k+1)} \leftarrow \frac{1}{\left|x_i^{(k)}\right| + \varepsilon}, \qquad i = 1, \ldots, n$
5: $\quad k \leftarrow k + 1$
6: **end while**

Output: $\mathbf{x}^* \leftarrow \mathbf{x}^{(k)}$

To demonstrate the effectiveness in CS-MRI the iterative weighting is applied on a well-known composite splitting technique, namely, the FCSA. As practised in any adaptive algorithm, the weights are updated in each iteration of the recursive ℓ_1 minimization algorithm. Although the number of operations per iteration would increase in the order of $O(n)$, but it gives better convergence and reconstruction compared to other reweighted algorithms as shown in Fig. 4.22.

Two different weighting vectors, namely, $\mathbf{W}^{\ell_1}$ and $\mathbf{W}^{TV}$ are required for solving the CS-MRI problem. $\mathbf{W}^{\ell_1}$ is used for the ℓ_1-norm and $\mathbf{W}^{TV}$ is used for the TV norm. Here, one can use either anisotropic or isotropic TV formulation. Considering the

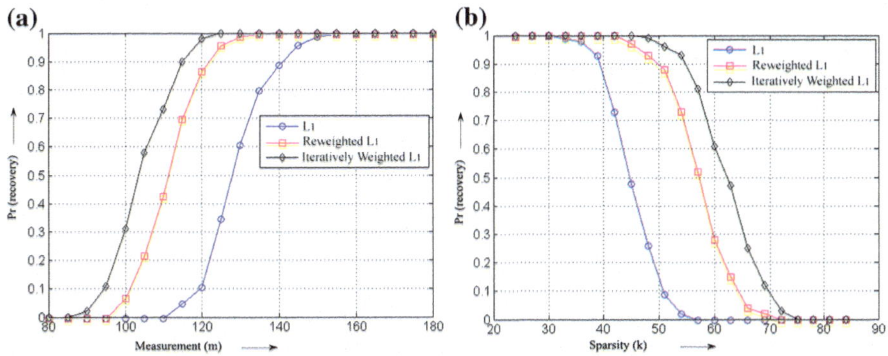

Fig. 4.22 Sparse signal recovery, **a** Fixed sparsity level, i.e., $k = 50$ with varying number of measurements and **b** Varying sparsity level with fixed number of measurements, i.e., $m = 120$

former for its simplicity, i.e., $||\mathbf{x}||_{TV} = \sum_i \left(|\mathbf{D}^{(1)}\mathbf{x}| + |\mathbf{D}^{(2)}\mathbf{x}|\right)$. Now, the weighted TV$\ell_1$-$\ell_2$ model can be defined as

$$\hat{\mathbf{x}} = \underset{\mathbf{x}}{\text{argmin}}\ \tfrac{1}{2}||\mathbf{F}_u\mathbf{x} - \mathbf{y}||_2^2 + \lambda_1||\ \mathbf{W}^{\ell_1}\Psi\mathbf{x}||_1\ + \lambda_2\ ||\mathbf{W}^{TV}\mathbf{D}\mathbf{x}||_1, \tag{4.3}$$

where $\mathbf{W}^{\ell_1}$ is a diagonal matrix and $\mathbf{W}^{TV}$ consists of two diagonal matrices as: $\mathbf{W}^{TV} = \begin{bmatrix} \mathbf{W}_x^{TV} \\ \mathbf{W}_y^{TV} \end{bmatrix}$. To solve the above weighted TV-ℓ_1-ℓ_2 model a procedure similar to that followed in the FCSA method is applied. For more details about the weighting scheme interested reader may refer [6, 9]. Algorithmic steps of iteratively weighted FCSA (IWFCSA) are shown in Algorithm 21. In algorithm, L is the Lipschitz constant.

Experiments are performed on an axial brain MR image and results are compared with the FCSA. Results in terms of PSNR and MSSIM are shown in shown in Table 4.2. From the table, it is observed that the reconstructed MR image using the IWFCSA shows an average improvement of 1.2 dB in PSNR compared to the FCSA. Similar conclusion is also drawn for MSSIM. For convergence, it is observed that the IWFCSA requires very less number of iterations compared to the FCSA to reach the same stopping criterion. For example, at 20% sampling ratio, the IWFCSA requires only 55 iterations whereas the FCSA requires 79 iterations. Similar results are also observed in terms of CPU Time.

Results in terms of number of iterations and CS measurements to achieve the same quality of reconstruction are shown in Tables 4.3 and 4.4. It is clearly observed IWFCSA requires significantly less number of iterations compared to the FCSA for giving similar visual results. For example, at 20% sampling ratio the IWFCSA requires only 21 iterations whereas the FCSA requires 79 iterations. It is also observed that the IWFCSA gives the same quality of reconstruction with significantly less number of k-space measurements. For example, the quality of reconstruction that would be achieved by the FCSA at 20% sampling ratio could be achieved with

Algorithm 21 Iteratively weighted fast composite splitting algorithm

Input: Ψ, $\mathbf{F}_u$, $\mathbf{y}$, λ_1, λ_2

Initialization: $\mathbf{x}^{(0)} \leftarrow \mathbf{F}_u^T\mathbf{y}$, $\left\{\mathbf{W}^{\ell_1},\ \mathbf{W}_x^{TV},\ \mathbf{W}_y^{TV}\right\} \leftarrow \mathbf{I}$, $t^{(1)} \leftarrow 1$, $\mathbf{r}^{(1)} \leftarrow \mathbf{x}^{(0)}$

while not converged **do**

$\mathbf{x}_g \leftarrow \mathbf{r}^{(k)} - L\nabla f\left(\mathbf{r}^{(k)}\right)$

$\mathbf{x}_1 \leftarrow \mathscr{T}\left(\mathbf{x}_g, \frac{\lambda_1}{L}\right)$

$\mathbf{x}_2 \leftarrow \mathscr{T}\left(\mathbf{x}_g, \frac{\lambda_2}{L}\right)$

$\mathbf{x}^{(k)} \leftarrow (\mathbf{x}_1 + \mathbf{x}_2)/2$

$\left(w^{\ell_1}\right)_{i,j}^{(k+1)} \leftarrow \frac{1}{\left|(\Psi x)_{i,j}^{(k)}\right| + \varepsilon_{l_1}}$

$\left(w^{TV}\right)_{i,j}^{(k+1)} \leftarrow \frac{1}{\left|(Dx)_{i,j}^{(k)}\right| + \varepsilon_{\mathrm{TV}}}$

$\mathbf{x}^{(k)} \leftarrow \text{project}\left(\mathbf{x}^{(k)}, [l, u]\right)$

$t^{(k+1)} \leftarrow \frac{\left(1+\sqrt{1+4(t^k)^2}\right)}{2}$

$\mathbf{r}^{(k+1)} \leftarrow \mathbf{x}^{(k)} + \frac{t^{(k)}-1}{t^{(k+1)}}\left(\mathbf{x}^{(k)} - \mathbf{x}^{(k-1)}\right)$

$k \leftarrow k+1$

end while

Output: $\mathbf{x}^* \leftarrow \mathbf{x}^{(k)}$

Table 4.2 Comparison of reconstruction results and convergence for various sampling ratios

SR (%)	FCSA				IWFCSA			
	PSNR (dB)	MSSIM	Itr.	CPU time (s)	PSNR (dB)	MSSIM	Itr.	CPU time (s)
10	31.91	0.9072	97	5.45	32.67	0.9261	62	4.33
15	33.85	0.9329	86	5.19	34.98	0.9516	59	4.21
20	35.91	0.9535	79	4.82	37.42	0.9688	55	4.10
25	38.74	0.9713	67	4.62	40.20	0.9801	32	2.25

Table 4.3 Comparison of required number of iterations for same quality of reconstruction result

SR (%)	FCSA		IWFCSA	
	PSNR (dB)	Itr.	PSNR (dB)	Itr.
10	31.91	97	31.95	20
15	33.85	86	33.91	20
20	35.91	79	36.03	21
25	38.74	67	38.82	21

17.5% sampling ratio in case of the IWFCSA, i.e., a reduction of 1638 k-space measurements. Figure 4.23 shows the reconstructed images obtained by the IWFCSA along with the FCSA. From visual inspection, it is observed that the former gives better reconstruction in terms preservation of contrast and edges which are indicated by white arrows in the figure.

Table 4.4 Comparison of number of measurements for same quality of reconstruction result

PSNR (dB)		Sampling ratio (%)		Number of measurements reduction
FCSA	IWFCSA	FCSA	IWFCSA	
31.91	32.04	10.0	08.5	983
33.85	33.89	15.0	13.0	1311
35.91	36.18	20.0	17.5	1638
38.74	38.74	25.0	22.0	1966

(a) (b) (c)

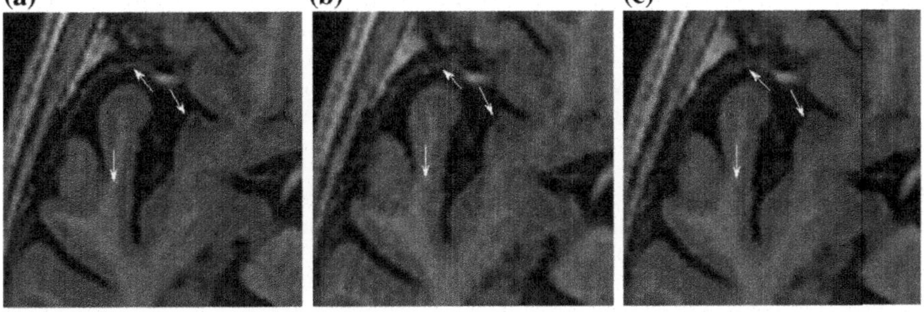

Fig. 4.23 Comparison of reconstructed image using the IWFCSA and the FCSA. **a** Original axial Brain image, **b** and **c** are reconstructed images using FCSA and IWFCSA, respectively

4.6 Conclusions

We have performed extensive experiments to show the performances of various CS reconstruction algorithms with a wide range of data sets. From the experimental results, we can conclude that among different categories composite splitting based algorithms are quite better for large-scale data. Moreover, we can also easily modify composite splitting based algorithms with additional sparsity enhancing regularization terms like structural sparsity, group sparsity, etc. Finally, in this chapter, we have also demonstrated the effectiveness of the iterative weighting scheme for CS-MRI reconstruction.

References

1. Afonso, M., Bioucas-Dias, J., Figueiredo, M.: Fast image recovery using variable splitting and constrained optimization. IEEE Trans. Image Process. **19**(9), 2345–2356 (2010)
2. Beck, A., Teboulle, M.: A fast iterative shrinkage-thresholding algorithm for linear inverse problems. SIAM J. Imaging Sci. **2**(1), 183–202 (2009)
3. Bioucas-Dias, J.M., Figueiredo, M.A.T.: A new TwIST: two-step iterative shrinkage/thresholding algorithms for image restoration. IEEE Trans. Image Process. **16**(12), 2992–3004 (2007)

4. Candes, E., Tao, T.: Near-optimal signal recovery from random projections: universal encoding strategies? IEEE Trans. Inf. Theory **52**(12), 5406–5425 (2006)
5. Candes, E., Wakin, M., Boyd, S.: Enhancing sparsity by reweighted L1 minimization. J. Fourier Anal. Appl. **14**(5), 877–905 (2008)
6. Datta, S., Deka, B.: Efficient adaptive weighted minimization for compressed sensing magnetic resonance image reconstruction. In: Proceedings of the Tenth Indian Conference on Computer Vision, Graphics and Image Processing, ICVGIP 16, pp. 95:1–95:8. ACM, New York (2016)
7. Daubechies, I., Defrise, M., De Mol, C.: An iterative thresholding algorithm for linear inverse problems with a sparsity constraint. Commun. Pure Appl. Math. **57**(11), 1413–1457 (2004)
8. Daubechies, I., Devore, R., Fornasier, M., Gunturk, C.: Iteratively reweighted least squares minimization for sparse recovery. Commun. Pure Appl. Math. **63**(1), 1–38 (2010)
9. Deka, B., Datta, S.: Weighted wavelet tree sparsity regularization for compressed sensing magnetic resonance image reconstruction. In: Advances in Electronics, Communication and Computing, Lecture Notes in Electrical Engineering, vol. 443, pp. 449–457. Springer, Singapore (2017)
10. Figueiredo, M., Nowak, R., Wright, S.: Gradient projection for sparse reconstruction: application to compressed sensing and other inverse problems. IEEE J. Sel. Top. Signal Process. **1**(4), 586–597 (2008)
11. Gorodnitsky, I.F., Rao, B.D.: Sparse signal reconstruction from limited data using FOCUSS: a reweighted minimum norm algorithm. IEEE Trans. Signal Process. **45**(3), 600–616 (1997)
12. Holland, P.W., Welsch, R.E.: Robust regression using iteratively reweighted least-squares. Commun. Stat. Theory Methods **6**(9), 813–827 (1977)
13. Huang, J., Zhang, S., Li, H., Metaxas, D.N.: Composite splitting algorithms for convex optimization. Comput. Vis. Image Underst. **115**(12), 1610–1622 (2011)
14. Huang, J., Zhang, S., Metaxas, D.N.: Efficient MR image reconstruction for compressed MR imaging. Med. Image Anal. **15**(5), 670–679 (2011)
15. Kim, S., Koh, K., Lustig, M., Boyd, S., Gorinevsky, D.: An interior-point method for largescale L_1-regularized least squares. IEEE J. Sel. Top. Signal Process. **1**(4), 606–617 (2008)
16. Lustig, M., Donoho, D., Pauly, J.M.: Sparse MRI: the application of compressed sensing for rapid MR imaging. Magn. Reson. Med. **58**, 1182–1195 (2007)
17. Ma, S., Yin, W., Zhang, Y., Chakraborty, A.: An efficient algorithm for compressed MR imaging using total variation and wavelets. In: IEEE Conference on Computer Vision and Pattern Recognition (CVPR 2008), pp. 1–8. Anchorage, AK (2008)
18. Wang, Z., Bovik, A.C., Sheikh, H.R., Simoncelli, E.P.: Image quality assessment: from error visibility to structural similarity. IEEE Trans. Image Process. **13**(4), 600–612 (2004)
19. Wright, S.J., Nowak, R.D., Figueiredo, M.A.T.: Sparse reconstruction by separable approximation. IEEE Trans. Signal Process. **57**(7), 2479–2493 (2009)
20. Yang, J., Zhang, Y., Yin, W.: A fast alternating direction method for TVL_1-L_2 signal reconstruction from partial Fourier data. IEEE J. Sel. Top. Signal Process. **4**(2), 288–297 (2010)

Chapter 5
CS-MRI Benchmarks and Current Trends

Abstract A significant progress has been already accomplished in compressed sensing magnetic resonance image reconstruction research. A few recent works have successfully integrated CS-MRI into the existing MRI scanner for clinical studies and within a short span of time it would be also available at a commercial scale. This chapter mainly aims to throw lights upon creating a set of common goals that practical CS-MRI reconstruction algorithms should project for successful implementation in medical diagnosis, and a few current research trends.

5.1 Introduction

Compressed sensing in magnetic resonance imaging (CS-MRI) technology possesses enough potential to enable rapid MRI into diagnostic imaging tool and thereby improve the patient comfort and healthcare economy. Rapid MRI is expected to maintain the diagnostic quality at par with industry standard. However, the trade-off is between the undersampling ratio and the quality of reconstruction.

Although CS reconstruction is an engineering problem, reconstructed images are supposed to be used for clinical diagnosis. So, rigorous performance evaluation by relevant experts are required. Smooth MR image looks visually better, but according to radiologists in many cases smoothing removes diagnostic details. Therefore, it is also required to follow medically standard methods/parameters for image quality evaluation.

MRI carry information of various pathological processes which is otherwise very difficult to acquire with a noninvasive process. It is safer than its close competitor for the same purpose, i.e., the computed tomography (CT) as the former is free from harmful radiations. Research in compressed sensing MRI has increased exponentially in recent times, but most of the approaches are in vitro. Only a few works are in the direction of diagnostic quality assessment; the underpinning requirement for compressed sensing MRI in clinical practice [17].

B. Deka and S. Datta, *Compressed Sensing Magnetic Resonance Image Reconstruction Algorithms*, Springer Series on Bio- and Neurosystems 9,
https://doi.org/10.1007/978-981-13-3597-6_5

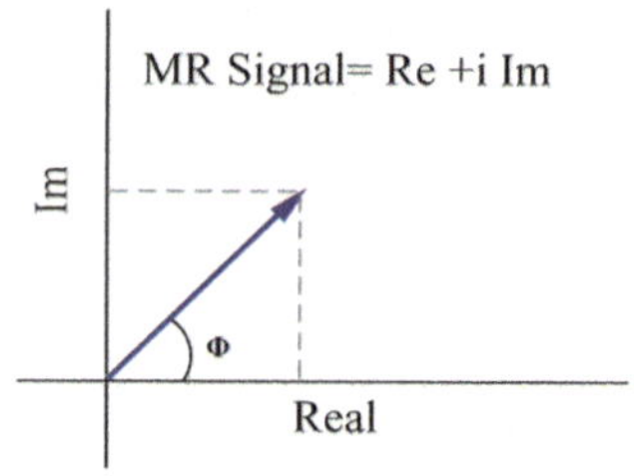

Fig. 5.1 Real and imaginary components of MR signal

5.2 Compressed Sensing for Clinical MRI

Literature study reveals that amount of data loss due to lossy image compression that may be tolerated in case of medical images basically depends on the extent of anatomical information present in the image, i.e., compression ratios of lossy DICOM images in clinical applications depend on the clinical relevance present in the image data. Going by the same philosophy, acceleration factor that can be achieved with the application of compressed sensing on clinical data depends on the underlying anatomical structures and pathological processes. So, in general, it is required to standardize the acceptance level of reconstructed MR image quality to fix the trade-off between diagnostic resolution and acceleration ratio.

An MR signal is detected using two receiver coils in orthogonal directions. The output of receiver coils are denoted by **I** (in phase) and **Q** (quadrature). Both signals contain equal amount of information, after post precessing they are combined to get the complex MR signal, i.e., MR signal(Re, Im) = Re + i Im as shown in Fig. 5.1, where $i^2 = -1$ imaginary unit. Now, from this complex data, we can easily obtain its magnitude and phase, i.e., Manitude $= \sqrt{Re^2 + Im^2}$ and Phase $(\Phi) = \tan^{-1}\left(\frac{Im}{Re}\right)$.

To evaluate the performance of different CS-MRI reconstruction techniques, simulations are carried out mainly in two ways. First, using magnitude MR images. It is the most commonly used approach because generally available images from the scanner are magnitude images, and in most cases radiologists prefer to use magnitude images only. Second, using raw k-space/complex MR data from the scanner. Although the magnitude images are better suited for visualization, the latter is more practical as raw data in frequency domain is already available and can be directly acquired using the CS principle. So, CS-MRI researchers prefer to use the second option for more realistic simulations. Figure 5.2 demonstrate detailed steps of the two CS-MRI data simulation techniques adopted by the CS-MRI research community.

5.3 CS-MRI Reconstruction

The quality of CS-MRI reconstruction mainly depends on three factors- (a) undersampling data acquisition, (b) sparsifying transformation, and (c) CS reconstruction technique.

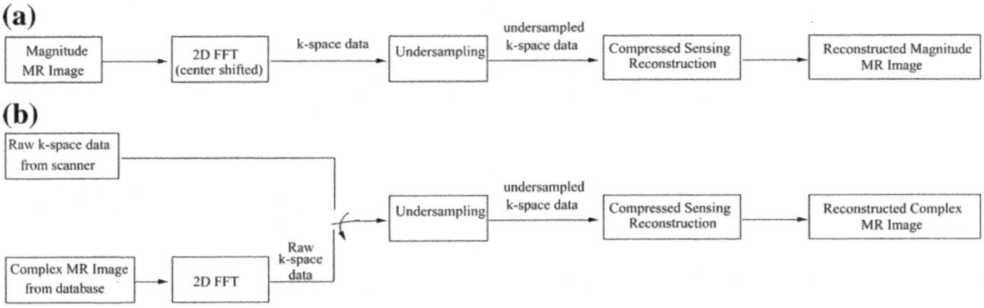

Fig. 5.2 Block diagram representation of compressed sensing MR image reconstruction. **a** Using magnitude MR image, **b** using raw k-space data or complex MR image

5.3.1 k-Space Undersampling in Practice and Sparsifying Transform

Number of k-space samples required to reconstruct a S-sparse signal is given by

$$m \geqslant C\mu^2 S \log(n) \tag{5.1}$$

where mutual coherence $\mu = \max\left|F_{\mathrm{u}}\Psi^{\mathrm{T}}\right|$ and signal length n are known. But, sparsity level S is unknown [22, p. 195]. Commonly MR images are compressible in a transform domain. For example, one can use wavelet transform to encode discontinuities present in the image. Wavelet transform of an image having simpler structures, like, cardiac image, produces relatively fewer high magnitude coefficients because there are only a few prominent edges in the image. On the other hand, in case of an image having complex structures, like, brain image has significant amount of edges and most of them are not prominent, therefore, only few coefficients have large magnitudes and most of the nonzero wavelet coefficients have low magnitudes. This means that S varies from image to image and it may be difficult to follow Eq. 5.1 in strict sense. However, in general, it is observed that in case of MRI it takes k-space samples roughly 2–5 times of the sparsity level S for good reconstruction. This would be approximately 20–30% of the full k-space data depending on the anatomical structure of the underlying image.

In conventional MRI, once the FOV of an image and resolution are fixed then the scan time taken by the machine depends on the minimum number of raw samples that must be acquired to fulfill the Nyquist criteria. In case of CS-MRI for clinical practice, undersampling strategy is one of the main issues. Randomized line undersampling trajectories in Cartesian coordinate system are relatively simpler and can be easily implemented in existing commercial scanners. Since most of the commercial systems are based on the Cartesian coordinate system, random undersampling in the phase direction can be conveniently done by modifying the MRI pulse sequence [20, p. 46].

Wavelet transform is the most widely used sparsifying transform in CS-MRI. For CS reconstruction, we need a measurement or sensing matrix which is maximally incoherent with the sparsifying transform. Since, real MRI data are in the frequency domain, so the measurement matrix that would be deployed for CS reconstruction can be simply an undersampling Fourier operator which is not implicit in the sensing of such signals. This measurement matrix is incoherent with wavelets at finer scales but relatively less incoherent at coarser scales. Due to this reason we need to use nonuniform undersampling at different scales of wavelet decomposition; more denser samples from lower frequencies (coarse scales) and sparser samples from higher frequencies (finer scales). This is also more realistic as in the k-space (frequency domain real MRI data), most of the energy is present near the center (low frequency region) and relatively less in the periphery (high frequency region). So to reduce the overall undersampling ratio it is obvious to acquire more dense samples from the center as compared to the periphery [22, p. 62].

As MR images are also piecewise smooth, first-order gradient of its intensity values exhibit sparsity in the spatial domain. This is mathematically modeled by the total-variation (TV) norm applied on the intensity image and then coupled with the wavelet sparsity as regularization terms to give rise to the most popular CS-MRI reconstruction model. Addition of TV norm with the translation variant discrete wavelet transform (DWT) would also help in restraining the artifacts during CS reconstruction due to the use of DWT alone. On the other hand, although the addition of more number of constraint improves reconstruction quality but it also increases the complexity of CS reconstruction [16, p. R307].

The above-mentioned sparsifying transforms, i.e., the DWT and TV as regularizing terms and variable density random line undersampling in Fourier domain are industry standards for CS-MRI reconstruction. Undersampling in Fourier domain is natural in MRI. For data acquisition using non-Fourier matrices, several research attempts have been already made but their practical implementations are not realizable [19, 37]. The 2D Variable density random undersampling pattern gives the best result in simulation but in practice it is not possible to implement for single-slice- or multi-slice MRI. It may be considered for acquisition of 3D volume data directly where random undersampling may be performed in both phase encode directions.

On the other hand, so far as sparsifying transforms are concerned, DWT is widely accepted for CS-MRI because it is sufficiently incoherent with the undersampling domain, i.e., Fourier domain, and wavelet coefficients of MR images are also sufficiently sparse. One can also consider other sparsifying transforms including the dual-tree complex wavelet transform and overcomplete contourlets or combinations of a few of them [28]. This makes the sparsifying transform more incoherent and more effective for CS reconstruction but at the cost of computational complexity and their clinical trials are yet to be checked. Another alternative is the use of learned overcomplete dictionaries for CS reconstruction. These dictionaries are adaptively learned from image patches extracted from some fully sampled reference images using the KSVD method [1]. This method is expected to give better reconstruction than wavelets at higher undersampling ratio due to the adaptive nature of the dictionary. Some works are already reported for CS-MRI reconstruction using learned dictionaries [3, 8, 25, 29]. They demonstrated improved CS reconstruction compared to fixed sparsifying transforms in terms of undersampling factor as well as quality of reconstruction but learning patch based dictionary requires fully sampled reference images besides the dictionary learning overhead. So, in a clinical setup they are somewhat questionable.

5.3.2 *Implementations*

CS-MRI reconstruction is an algorithmically controlled rapid MRI tool that works in association with the existing scanners; can be broadly classified into two categories, namely, the software based approach through simulations using software tools, like, MATLAB, SCILAB, and the hardware-based approach using embedded C/C++ or CUDA. The former focusses to develop new algorithms tailored for CS-MRI reconstruction and evaluate their effectiveness with the existing ones through convergence analysis, performance evaluation, etc. On the other hand, the latter put emphasis on physical realization of CS-MRI in clinical practice.

CS-MRI is a rapidly growing field, and over less than a decade new algorithms are continuously coming from different research groups. However, only recently considerable hardware based works are being implemented and a few have already been successfully integrated in clinical environments [21, 33]. More hardware works are still awaited to make CS-MRI scanners commercially abundant. The standard reconstruction model is as follows:

$$\hat{\mathbf{x}} = \underset{\mathbf{x}}{\operatorname{argmin}} \, ||\mathbf{F}_u\mathbf{x} - \mathbf{y}||_2^2 + \lambda_1 ||\Psi\mathbf{x}||_1 + \lambda_2 \, ||\mathbf{x}||_{TV}$$

where $\mathbf{F}_u$ is the undersampling Fourier transform and Ψ is the sparsifying transform. Recently, a number of modifications of the above model are being done to exploit intra-slice and inter-slice redundancy by incorporating group sparsity or structural sparsity. Among them a few are as follows: (a) besides standard wavelet and gradient sparsities, another sparsity promoting regularization term is introduced as a constraint [6]. It models wavelet coefficients as overlapping parent–child groups in a quadtree structure. The new problem, *Wavelet Tree Sparsity MRI (WaTMRI)*, may be solved efficiently using a hybrid splitting approach discussed in the previous chapter, (b) concept of parent–child grouping of detailed wavelet coefficients may be extended to multichannel images since they have similar edges and structures across channels; parent–child groups belonging to different trees may be grouped together as forest, [7] so that coefficients of different channel images have similar patterns (large or small) at the same position across multiple channels. This arrangement of multichannel wavelet coefficients enforce a particular forest group to be zero or nonzero according to the ℓ_2-norm value of the respective group, and (c) spatial domain finite differences in a particular spatial location from multiple slices may be grouped together because in multi-slice sequence adjacent slices follow a similar edge pattern.

Let $\mathbf{Y} = \left[\mathbf{y}_1, \mathbf{y}_2, \ldots, \mathbf{y}_t\right]$ be the 3D undersampled data, and $\mathbf{X} = [\mathbf{x}_1, \mathbf{x}_2, \ldots, \mathbf{x}_t]$, the corresponding data in spatial domain, then the CS-MRI reconstruction model for multi-slice data may be defined as-

$$\hat{\mathbf{X}} = \arg\min_{\mathbf{X}} \frac{1}{2} \sum_{l=1}^{t} ||\mathbf{F}_u \mathbf{X}_l - \mathbf{Y}_l||_2^2 + \lambda_1 \, ||\mathbf{X}||_{\mathrm{JTV}} + \lambda_2 \sum_{g \in \mathbf{G}} \mathbf{W}_G || \left(\Psi \mathbf{X}_{l=1,\ldots,t}\right)_g ||_2$$

where t indicates the number of slices in joint reconstruction and $\mathbf{W}_G$ is a weighting matrix. This may be considered as the state-of-the-art reconstruction model for multi-slice MRI. It is also applicable for multichannel and multi-contrast CS-MRI [13].

Murphey et al. [24] implemented the ℓ_1-SPRiT POCS algorithm using multi-core CPU and GP-GPU architectures for 3D multichannel MRI. To speedup the reconstruction computationally expensive operations, like, wavelet and Fourier transforms are implemented in parallel. In [5], the IST algorithm was implemented with parallel architectures with GP-GPU and similarly in [4], the OMP algorithm was implemented with GP-GPU. Sabbagh et al. [30] implemented ℓ_1-ESPIRiT CS-MRI reconstruction technique using the Berkeley Advanced Reconstruction Toolbox (BART) [34]. Recently, Schaetz et al. [31] implemented nonlinear conjugate gradient algorithm on a Super-micro Super-Server 4027GR-TR system.

5.4 Image Quality Assessment

Clinical demand of MRI is continuously increasing due to its ability to provide high contrast soft tissue images, availability for imaging in any arbitrary plane and development of technology to reduce scan time. In medical imaging, quality is very important for subsequent diagnosis. In MRI, quality of images may vary with sites where they are acquired due to ambient variations, operator skill, hardware variations, etc. Thus quality of images in one site may not match with those from another site even with the same parameter setting. Therefore, it is very important to standardize acceptable levels of image quality.

In image quality evaluation objective quality measurement parameters are most preferred. Because, these indices make computer-aided medical image analysis easier. Moreover, these results are reproducible because they are calculated by well-defined standard algorithms [32]. Signal-to-noise ratio (SNR) and contrast-to-noise ratio (CNR) are the most commonly used region of interest (ROI) based image quality evaluation metrics as area of interest need to be unaffected during image compression or transmission.

According to Aja-Fernandez et al. [2] structural information of any image is encoded in the distribution of its local variance. Based on this theory they have proposed a new image quality evaluation metric, namely, quality index based on local variance (QILV). They have validated the performance of the QILV with MR images affected by Rician noise. It is observed that the QILV is not only able to indicate degradation of image quality due to noise but also detect blurring due to filtering.

The most widely used clinical full-reference-based image quality evaluation metric is root mean square error (RMSE). Computationally it is quite simple and numerically it indicates the deviation from reference image. But it is observed that two images having same RMSE value may be visually very much dissimilar.

In medical image analysis, quality evaluation metric should be based on the characteristic of human visual system (HVS). Because finally medical images are inspected and interpreted by human. HVS is very complex, although a sufficient amount of research is carried out on HVS, it is not enough to exactly model it. Number of attempts have been taken to develop the quality evaluation metric based on HVS. One of the well-known HVS-based image quality metric is the mean structural similarity (MSSIM) index [36]. It is based on the idea that human eyes are more sensitive to the structural information. MSSIM gives a scaler quantity in between 0 to 1 by comparing the structural similarity between reference image and test image. A scalar quantity close to 1 indicates that structural information is well preserved in respect to the reference image. In literature, a number of works utilize the MSSIM index for evaluation of MR image quality [10, 11, 15, 32].

Prieto et al. [27] proposed a subjective MR image quality evaluation technique based on just noticeable differences (JND) [9], namely, JND scanning (JNDS). There is a maximum threshold level below which distortions in any pixel is indistinguishable at a particular gray level and contrast by human eyes. JND is a binary image where positions of one's indicate locations of pixels where difference of two images is noticeable. In [15, 27], authors demonstrate how difference between two images disappear as contrast level decreases. If we continuously decrease the contrast level of both images, the pixel for which JND disappear initially is the least distorted pixel and pixel for which JND disappear towards the ending of the process is the most distorted pixel. After summing the information for all contrast levels the quotient of number of pixels having probability one to the number of pixels having probability zero is defined as JNDS index.

Subjective image quality evaluation, like, mean opinion score (MOS) is often used as a benchmark to validate any HVS based image quality metric [11, 32]. However, subjective visual inspection depends on level of experience of observer and results may not be reproducible. As it is based on HVS characteristics, so it may be extended to the perception of radiologists.

It may be also very good idea to compare MOS with the JNDS to see that subjective quality assessment follows the quantitative assessment [15, 27] linearly.

5.5 Computational Complexity

In CS-MRI reconstruction fast forward/ reverse Fourier and wavelet transformations are the most computationally expensive operations which is $O(n \log n)$ for 2D images. Thus, the computational complexity of CS-MRI reconstruction algorithms are O (n log n). However, overall computational costs and reconstruction times vary depending on the algorithms.

CS-MRI reconstruction algorithms are generally sequential in nature. But for utilization of advanced computational resources in both multi-core CPU and GP-GPU architectures parallel implementations are necessary. In [31, Table 5], authors demonstrate the change in reconstruction speed with increasing number of GPUs and reveal that after a particular level if number of GPUs are increased then there will be no further acceleration in the reconstruction. This means that acceleration of CS-MRI reconstruction speed is not possible just by increasing the computing resources only; we need to explore more on finding solutions for optimal parallel implementations of the relevant algorithm.

5.6 Current Trends

5.6.1 Interpolated CS-MRI (iCS-MRI) Reconstruction

For analysis of diagnostic details of human organs 3D or 2D multi-slice MRI is most commonly carried out in clinics. Modern MRI scanner improves the data acquisition time by increasing number of parallel receiver coils, coil orientations, etc. In spite of this, data acquisition is quite slow for 3D or higher dimensional applications where multi-slice data acquisition is very common; adjacent slices being highly correlated as they maintain either zero or very small inter-slice gaps, and 100–300 such slices are to be acquired for completing a single 3D scan.

To utilize inter-slice similarity, Pang and Zhang [26] introduce the idea of interpolated compressed sensing (iCS). As the adjacent slices are highly correlated, iCS technique follows a nonuniform undersampling scheme across slices. Some slice are highly undersampled (H-slice) and some slices are lightly undersampled (L-slice). Missing samples of any particular H-slice is interpolated from an adjacent L-slice. After interpolation of all H-slices, each H-slice is independently reconstructed using a CS-MRI technique. This technique significantly reduces the amount of overall undersampling ratio in 2D multi-slice MRI.

Recently, authors in [14] proposed an efficient iCS technique. They have reduced the computational cost of interpolation to $O(n)$ from $O(n \log n)$ in [12] and $O(n^2)$ in [26]. They also experimentally demonstrate that their interpolation scheme is significantly faster than [12, 26]. The idea of iCS may be extended to multichannel MRI as adjacent channel images are highly correlated.

5.6.2 Fast CS-MRI Hardware Implementation

The state-of-the-art MRI scanners are equipped with multiple receiver coils working in parallel; each coil collects only a part of the full k-space data. Most recent systems are able to collect as many as 32 channels of data in parallel [35]. CS-MRI reconstruction may be integrated with the existing parallel scanners without changing the coil configuration. However, it is computationally expensive when we consider CS with parallel MRI.

Only a few CS-MRI works have been implemented with clinical settings. For clinical practice, any CS-MRI reconstruction technique should not take more than 2 minutes [21]. It is because immediate feedback is necessary to decide whether re-examination is required or not for that particular field-of-view. Dr. Shreyas Vasanawala, a radiologist and his CS-MRI research group translated CS-MRI research into new technology for medical imaging. They integrated CS-MRI reconstruction hardware with existing scanners at Lucile Packard Children's Hospital, Stanford. Mann et al. [23] achieved 3.8 times acceleration in fat fraction mapping of human liver without degrading diagnostic image quality using compressed sensing. It shortens the breath hold period to 4.7 from 17.7 s in a 3T scanner. Recently, Toledano–Massiah et al. [33] implemented CS in clinical MRI at Fondation H$\hat{o}$pital Saint Joseph by integrating the CS technology with conventional 3D-FLAIR sequence and achieved acceleration factor of 1.3.

System memory and data transfer are major limitations for 3D and dynamic MRI with higher number of coils. For example, dynamic cardiac MRI with $(256 \times 256 \times 128)$ spatial matrix $\times 24$ times points $\times 16$ channels require 24 GB memory space just to store the complex data matrix. Similarly, blood flow imaging also suffer from low resolution, long scan time and huge amount of data. To process these huge amounts of data require parallel processing systems with large memory capacity. The development of parallel processors increased the computational throughput for data intensive operations. The computational throughput of the processing unit is directly proportional to the number of cores present in it. Intel and AMD can provide CPU with 4–16 cores per socket.

Murphy et al. [24] presents a parallel CS-MRI reconstruction with clinically feasible time via multi-core CPUs and GPUs. MRI reconstruction techniques have nested data parallelism. Computationally expensive operations, like, Fourier and wavelet transformations are performed in parallel for multichannel multi-slice data. With advanced processor architectures one can exploit four levels of parallelism. For example, in multichannel multi-slice MRI reconstruction, each operation is performed over 4D matrix representing the multichannel 3D data. For processing this type of volumetric data one can use two-level parallelism across slices and channels but require frequent synchronization. Moreover, 3D reconstruction problem for each slice can be decoupled into multiple independent 2D problems which do not require any synchronization. It provides efficient parallelism of volumetric MRI reconstruction; solves multiple 2D reconstructions per GPU in batch mode, exhibits better parallelism and achieves efficient resource utilization.

Synchronization at higher levels of processing hierarchy is computationally expensive. If image matrix size is large or less number of cores are available in CPU then one can exploit slice wise parallelism at higher levels of processing hierarchy. On the other hand if image matrix size is small or large number of cores are available in CPU then one should exploit parallelism in channel wise and decouple

3D problems into 2D problems. In future, it is expected that number of cores and amount of cache memory per core will increase over time which will make the latter more realistic.

A few fast CS-MRI hardware realizations in reverse chronological order are

1. Schaetz et al. [31] implemented nonlinear conjugate gradient algorithm on a server system equipped with PCIe 3.0, 2x Intel Xeon Ivy Bridge-EP E5-2650 processors, 128 GB RAM and 8× NVIDIA Kepler GK110 GPU having 6 GB memory each. They could reconstruct dynamic cardiac images of size 160×160 at the rate of 22 frames per second.
2. Sabbagh et al. [30] implemented ℓ_1-ESPIRiT CS-MRI reconstruction technique using the Berkeley Advanced Reconstruction Toolbox (BART) [34] on a workstation equipped with dual Intel E5 2650 CPUs, 128 GB of memory and NVIDIA Tesla K20m GPU with 2496 computing cores. They have reconstructed $256 \times 128 \times 113$ volume data acquired with 6-fold undersampling ratio in 131 s.
3. Murphy et al. [24] reconstructed 8 and 16 channels $192 \times 256 \times 58$ volume data in 20 and 35 s respectively, and 32 channel same data matrix in about 80 s using the ℓ_1-SPRiT algorithm implemented with a system having 2x 6-core Intel Xeon X5650, 2.67GHz processors with 64GB DRAM and four NVIDIA GTX580 GPU cards with 3GB DRAM in each.
4. Kim et al. [18] demonstrated the CS-MRI reconstruction of 8-channel $256 \times 160 \times 80$ volume data in a computer equipped with Intel Core i7 processor and NVIDIA Tesla C2050 GPU using modified SENSE algorithm in 19 s.
5. Vasanawala et al. [35] share their two years of experience in practical CS-MRI reconstruction for accelerating pediatric body MRI. They have successfully achieved reconstruction in 120.7 s for $192 \times 320 \times 110 \times 32$ data using the POCS algorithm in an Intel Westmere system with 2x 6-core 2.67GHz processors and 4 NVIDIA Tesla C1060 GPUs.

5.7 Future Research Directions

Some promising future research directions in CS-MRI reconstruction-

- Improvement of the quality of reconstruction by adding more sparsity promoting terms in the CS reconstruction model.
- Development of MR image model using the probability theory and adding them as regularization priors for further improvement of the quality of CS reconstruction.
- Development of optimal under sampling patterns and sparsifying transforms for clinically feasible implementations.
- Implementation of parallel CS reconstruction hardware using GP-GPUs for 3D or higher dimensional MRI data at diagnostic resolutions and within the clinically feasible time.

5.8 Conclusions

In this chapter, authors made an attempt to set benchmarks for CS-MRI reconstruction in both software and hardware platforms. Some important recent trends in this topic are also briefly discussed with mentions to a few promising future research directions. Although relatively less number of works are carried out in the direction of clinically feasible implementation, but some CS-MRI research organizations successfully implemented and integrated CS for clinical practice, for example, Lucile Packard Children's Hospital, Stanford and Fondation Hôpital Saint Joseph, Paris. It is also expected that in near future all clinical MRI scanners will be integrated with the CS-MRI technology.

References

1. Aharon, M., Elad, M., Bruckstein, A.: k-SVD: an algorithm for designing overcomplete dictionaries for sparse representation. IEEE Trans. Signal Process. **54**(11), 4311–4322 (2006)
2. Aja-Fernandez, S., San Jose Estepar, R., Alberola Lopez, C., Westin, C.F.: Image quality assessment based on local variance. In: 28th IEEE EMBS, pp. 4815–4818. New York City, USA (2006)
3. Bilgin, A., Kim, Y., Liu, F., Nadar, M.S.: Dictionary design for compressed sensing MRI. Proc. Intl. Soc. Mag. Reson. **Med**, 4887 (2010)
4. Blanchard, J.D., Tanner, J.: GPU accelerated greedy algorithms for compressed sensing. Math. Program. Comput. **5**(3), 267–304 (2013)
5. Borghi, A., Darbon, J., Peyronnet, S., Chan, T.F., Osher, S.: A simple compressive sensing algorithm for parallel many-core architectures. J. Signal Process. Syst. **71**(1), 1–20 (2013)
6. Chen, C., Huang, J.: Exploiting the wavelet structure in compressed sensing MRI. Magn. Reson. Imaging **32**, 1377–1389 (2014)
7. Chen, C., Li, Y., Huang, J.: Forest sparsity for multi-channel compressive sensing. IEEE Trans. Signal Process. **62**(11), 2803–2813 (2014)
8. Chen, Y., Ye, X., Huang, F.: A novel method and fast algorithm for MR image reconstruction with significantly under-sampled data. Inverse Probl. Imaging **4**, 223–240 (2010)
9. Chou, C.H., Li, Y.C.: A perceptually tuned subband image coder based on the measure of just-noticeable-distortion profile. IEEE Trans. Circuits Syst. Video Technol. **5**(6), 467–476 (1995)
10. Chow, L.S., Paramesran, R.: Review of medical image quality assessment. Biomed. Signal Process. Control. **27**, 145–154 (2016)
11. Chow, L.S., Rajagopal, H., Paramesran, R.: Correlation between subjective and objective assessment of magnetic resonance MR images. Magn. Reson. Imaging **34**(6), 820–831 (2016)
12. Datta, S., Deka, B.: Magnetic resonance image reconstruction using fast interpolated compressed sensing. J. Opt., 1–12 (2017)
13. Datta, S., Deka, B.: Multi-channel, multi-slice, and multi-contrast compressed sensing MRI using weighted forest sparsity and joint TV regularization priors. In: 7th International Conference on Soft Computing for Problem Solving (SocProS) (2017)
14. Datta, S., Deka, B.: An efficient interpolated compressed sensing reconstruction scheme for 3D MRI (2018). Manuscript submitted for publication
15. Deka, B., Datta, S., Handique, S.: Wavelet tree support detection for compressed sensing MRI reconstruction. IEEE Signal Process. Lett. **25**(5), 730–734 (2018)
16. Hollingsworth, K.G.: Reducing acquisition time in clinical MRI by data undersampling and compressed sensing reconstruction. Phys. Med. Biol. **60**(21), R297 (2015)

17. Jaspan, O., Fleysher, R., Lipton, M.: Compressed sensing MRI: A review of the clinical literature. Br. J. Radiol. **88**(1056), 1–12 (2015)
18. Kim, D., Trzasko, J., Smelyanskiy, M., Haider, C., Dubey, P., Manduca, A.: High-performance 3D compressive sensing MRI reconstruction using many-core architectures. Int. J. Biomed. Imaging **2011**, 1–11 (2011)
19. Liang, D., Xu, G., Wang, H., King, K.F., Xu, D., Ying, L.: Toeplitz random encoding MR imaging using compressed sensing. IEEE ISBI **2009**, 270–273 (2009)
20. Lustig, M.: Sparse MRI. Ph.D. thesis, Electrical Engineering, Stanford University (2008)
21. Lustig, M., Keutzer, K., V.S.: The Berkeley Par Lab: progress in the parallel computing landscape, chap. In: Introduction to parallelizing compressed sensing magnetic resonance imaging, pp. 105–139. Microsoft Corporation (2013)
22. Majumdar, A.: Compressed Sensing for Magnetic Resonance Image Reconstruction. Cambridge University Press, New York (2015)
23. Mann, L.W., Higgins, D.M., Peters, C.N., Cassidy, S., Hodson, K.K., Coombs, A., Taylo, R., Hollingsworth, K.G.: Accelerating MR imaging liver steatosis measurement using combined compressed sensing and parallel imaging: a quantitative evaluation. Radiology **278**(1), 247–256 (2016)
24. Murphey, M., Alley, M., Demmel, J., Keutzer, K., Vasanawala, S., Lustig, M.: Fast ℓ_1-SPIRiT compressed sensing parralel imaging MRI: Scalable parallel implementation and clinically feasible runtime. IEEE Trans. Med. Imaging **31**(6), 1250–1262 (2012)
25. Otazo, R., Sodickson, D.K.: Adaptive compressed sensing MRI. In: Proceedings of ISMRM, p. 4867. (2010)
26. Pang, Y., Zhang, X.: Interpolated compressed sensing for 2D multiple slice fast MR imaging. Ed. Jonathan A. Coles. PLoS ONE **8**(2), 1–5 (2013)
27. Prieto, F., Guarini, M., Tejos, C., Irarrazaval, P.: Metrics for quantifying the quality of MR images. In: Proceedings of 17th Annual Meeting ISMRM, vol. 17, p. 4696 (2009)
28. Qu, X., Cao, X., Guo, D., Hu, C., Chen, Z.: Combined sparsifying transforms for compressed sensing mri. Electron. Lett. **46**(2), 121–123 (2010)
29. Ravishankar, S., Bresler, Y.: Mr image reconstruction from highly undersampled k-space data by dictionary learning. IEEE Trans. Med. Imaging **30**(5), 1028–1041 (2011)
30. Sabbagh, M., Uecker, M., Powell, A.J, Leeser, M., Moghari, M.H.: Cardiac MRI compressed sensing image reconstruction with a graphics processing unit. In: 2016 10th International Symposium on Medical Information and Communication Technology (ISMICT), pp. 1–5 (2016)
31. Schaetz, S., Voit, D., Frahm, J., Uecker, M.: Accelerated computing in magnetic resonance imaging: Real-time imaging using nonlinear inverse reconstruction. Comput. Math. Methods Med. **2017**, 1–11 (2017)
32. Sinha, N., Ramakrishnan, A.: Quality assessment in magnetic resonance images. Crit. Rev. Biomed. Eng. **38**, 127–141 (2010)
33. Toledano-Massiah, S., Sayadi, A., de Boer, R.A., Gelderblom, J., Mahdjoub, R., Gerber, S., Zuber, M., Zins, M., Hodel, J.: Accuracy of the compressed sensing accelerated 3d-flair sequence for the detection of ms plaques at 3t. AJNR. Am. J. Neuroradiol., 1–5 (2018)
34. Uecker, M., Ong, F., Tamir, J.I., Bahri, D., Virtue, P., Cheng, J.Y., Zhang, T., Lustig, M.: Berkeley advanced reconstruction toolbox. Proc. Intl. Soc. Mag. Reson. Med. **23**, 2486 (2015)
35. Vasanawala, S., Murphy, M., Alley, M., Lai, P., Keutzer, K., Pauly, J., Lustig, M.: Practical parallel imaging compressed sensing MRI: Summary of two years of experience in accelerating body MRI of pediatric patients. In: IEEE International Symposium on Biomedical Imaging: From Nano to Macro 2011, pp. 1039–1043. Chicago, IL (2011)
36. Wang, Z., Bovik, A.C., Sheikh, H.R., Simoncelli, E.P.: Image quality assessment: from error visibility to structural similarity. IEEE Trans. Image Process. **13**(4), 600–612 (2004)
37. Yin, W., Morgan, S., Yang, J., Zhang, Y.: Practical compressive sensing with toeplitz and circulant matrices. In: Proc. SPIE Vis. Commun. Image Process. **7744**, 1–10 (2010)

Chapter 6
Applications of CS-MRI in Bioinformatics and Neuroinformatics

Abstract MRI has a number of applications in bioinformatics and neuroinformatis, like, functional MRI (fMRI), diffusion weighted MRI (DW-MRI), and magnetic resonance spectroscopy (MRS). It gives valuable information about anatomical structure, the functioning of organs, neuronal activity, and abnormality inside the human body. Although MRI has a number of clinical advantages, it suffers from a fundamental limitation, i.e., slow data acquisition resulting in low SNR, poor resolution, and patient discomfort. Applications of CS-MRI for clinical practice is only a few till date but this new technology has a tremendous potential to overcome the fundamental limit of conventional MRI.

6.1 Introduction

Bioinformatics deals with the analysis and modeling of biological data using machine learning. It brings together biological science, computer science, mathematics, and engineering for understanding of biological system through modeling and simulation [20, p. 7]. It requires concepts and expertise from computer science, mathematics, physics, and engineering besides the biological science. Modern medical imaging modalities helps in modeling and analysis of anatomical details, like, tissue properties, structure, surface areas, etc. Then this information may be used for analysis, and diagnosis of biological system and may also be used for testing and validation of already developed computational models [21].

Digital image processing techniques, like, image segmentation and image registration significantly contribute for modeling of system biology. For example, segmentation helps to extract an area where a biological activity of interest has occurred. Registration helps in alignment of image sequences collected at different times. Advanced imaging and visualization technology and graphics workstation becomes a core part of any modern medical imaging equipment.

Medical imaging plays a key role in diagnosis [12]. Images obtained using different imaging techniques may look similar, like, CT and MRI. However, they contain different clinical information and underlying physics for mapping physiological parameters besides medical usefulness. Conventionally, CT and MRI are used to

B. Deka and S. Datta, *Compressed Sensing Magnetic Resonance Image Reconstruction Algorithms*, Springer Series on Bio- and Neurosystems 9,
https://doi.org/10.1007/978-981-13-3597-6_6

produce 3D volumes for clinical study of human physiology and its well-being. For proper and correct diagnosis, understanding of abnormalities in physiological functioning is more important than the anatomical details, for which functional and dynamic MRI play a promising role. Development in imaging technology also emphasizes noninvasive interventions and treatments, drug delivery, surgical planning, and simulation.

In modern healthcare system, applications of medical images are not limited to clinical diagnosis only, they are also used for transmission and storage through picture archiving and communication systems (PACS) where images from multiple modalities may be transmitted to a remote location. Due to availability of images in digital form and highly equipped computer and networks, images are now considered as core for biomedical informatics. It involves image acquisition, management of images, image representation, and interpretation [37].

Physician and Biologist collect data from human body or conduct experiment using biological samples to find unknown facts. This information can be used for modeling and inference of biological processes with the help of Bioinformatics. For example, unleashing the evolution of cancer stem cells (CSCs) would require modeling and simulations of CSCs which would play a crucial role in the development of CSC-targeted anticancer therapies in future [20, ch. 28].

Human brain is considered as one of the most complex information processing system. The relevant scientific subject which aims to understand brain and nervous system functioning and creating equivalent intelligent systems is known as neuroinformatics [20, p. 9]. It involves processing, analysis and modeling of information obtained from brain and nervous system. Human brain has pre-allocated areas for processing of each type of information like, language, visual, audio, etc. Neuron or nerve cell is the primary element of our central nervous system (CNS). A neuron receives, processes, and transmits biological information in the form of electrical or chemical signal. A single neuron can be connected to thousands of neurons via synapses. Functionally there are different types of neurons, like, motor neurons, sensory neurons, etc.

Neuroinformatics brings together research and developments of neuroscience and informatics. The main target of neuroinformatics is to understand the functionality of human nervous system. Conceptually, it can be divided into four major areas- (a) *Neuroscience Knowledge Management*, it is an application of computer science and information technology in neuroscience for managing knowledge, database representation and architecture, interoperablity. (b) *Computational Modeling and Simulation*, it involves modeling and simulations, and different approaches of data mining. (c) *Imaging*, it aims on data representation and structural complexity of human brain using noninvasive imaging techniques, like, functional MRI (fMRI). (d) *Genetics and Neurodegeneretive Diseases*, it involves genomic approaches to analysis the human nervous system and its abnormalities [6].

Neuroimaging is a technique of acquiring structural and functional information of human nervous system through imaging. Physicians who are specialized in this area are called neuroradiologist. Neuroimaging can be classified into two broad categories, namely, *structural imaging* and *functional imaging*. Former one involves with the gross structure of human nervous system and its diagnosis for intracranial disease, tumor, or injury. Latter one related to functional activity of nervous system and diagnosis of metabolic disease, cognitive psychology, and brain–computer interface [39].

MRI is not only able to show anatomical structures but also functioning of tissues and organs in the human body. Structural MRI is more commonly used in clinical practice, includes T1, T2, PD and diffusion weighted [34]. T1 and T2 weighted sequences are the core of clinical MRI. While T1-weighted images are generally used to study normal anatomical structure, but T2-weighted images are used for the detection of abnormality, like edema. Recently, fluid attenuated inversion recovery (FLAIR) sequence has replaced the conventional T2-weighted sequence as it suppresses the CSF signal and increases the lesion to background CSF contrast. Moreover, it reduces the acquisition time and artifacts [39]. PD weighted images are commonly used to detect Meniscus tears in the knee. It is also used in brain MRI to detect abnormality in gray or white matter. Diffusion weighted MRI (DW-MRI) maps the diffusion process of water molecules in a tissue. Intensity of DW-MRI indicates the rate of diffusion of water molecules on a particular field-of-view (FOV). It has a significant role in clinical practice for early detection of ischemic strokes and to distinguish acute strokes from mild strokes.

On the other hand, functional MRI (fMRI) includes blood oxygen level dependent (BOLD) and perfusion imaging. BOLD fMRI identify active areas of brain at the time of data acquisition by detecting the oxygenated blood level. It is commonly used for tracking neuronal activities. Perfusion imaging uses perfusion tracer to produce differential contrasts in tissues. It is commonly used for the measurement of cerebral blood flow (CBF). Magnetic resonance angiography (MRA) is another example of fMRI. It determines how well blood vessels are working. It is very commonly used to look at the arteries of neck, brain, heart, lungs, kidneys, and legs.

Magnetic resonance spectroscopy (MRS) and chemical shift imaging are different from above categories of MRI, they measure chemical and metabolic changes that occur due to tumors or other disorders [39].

6.2 MRI in Bioinformatics

In body MRI entire body or any part is imaged with one or more sequences for analysis and diagnosis of multi-organ diseases. Due to larger FOV acquired images suffer from low SNR and resolution. According to the CS-MRI theory, body MRI seems to be one of the most favorable applications. It has significant potential to reduce drawbacks associated with body MRI. To evaluate its clinical effectiveness, some research groups have already implemented and integrated CS-MRI into traditional scanners for clinical purpose, for example, at Lucile Packard Children's Hospital, Stanford for breath-hold cardiac imaging of pediatric patients [25] and Fondation Hôpital Saint Joseph, Paris for detection of multiple sclerosis plaques using 3D FLAIR sequence [42]. Recently, manufacturers have also added CS-based MRI scanners in their product list [10, 22].

6.2.1 Whole-Body MRI

Analysis of some diseases require imaging of multiple organs and regions of the human body. Pulse sequences commonly used are (a) STIR, (b) T1-weighted fast spin echo, (c) Contrast-enhanced T1-weighted 3D gradient echo, (d) Single-shot fast spin echo, and (e) Steady-state free precession [4]. Due to the technological advancement it is possible to acquire MR image of the whole body. In whole-body MRI, entire body is scanned in multiple planes with multiple imaging sequences. It provides anatomical details of the whole body but does not give any functional information [4]. Initially it is used for only children to identify the stage of lymphoma with short inversion time inversion recovery (STIR) sequence. Later on its use is expanded to both children as well as adult for cancer staging and other diseases.

MRI is often used for breast imaging for patients already diagnosed with possible breast cancer for reconfirmation and women with genetic predisposition [11]. It is possible to detect some cancers with MRI which are not easily identified on mammograms. T1-weighted gradient echo and short T1 inversion recovery (STIR) sequence are commonly used for breast MRI. Conventional MRI can detect cancer when it becomes a few millimeters or millions of cells. But, for cancer cure early detection and treatment are very essential. Recent developments in nanotechnology provides nanomaterials for early cancer detection using MRI [3].

Main limitation in the whole-body MRI is the acquisition time, as FOV is large it takes significantly longer scan time. Moreover, conventional whole-body MRI suffers from low SNR and low resolution artifacts. CS in clinical MRI would be able to overcome these drawbacks. Because with the help of this new technology one can reconstruct clinically acceptable image/volume from just 20–30% of the full k-space data depending on the underlying anatomical details.

Siemens Healthineers in 2016 at the Annual Meeting of the Radiological Society of North America (RSNA) in Chicago, USA introduces the CS-MRI technology. CS-MRI can be performed within a fraction of time that any conventional MRI scan requires. Dr. Christoph Zindel, Vice President of Magnetic Resonance at Siemens Healthineers states “Compressed Sensing enables scanning speeds that we could only dream of before”. With the help of CS technology it is possible to reduce MRI scan time up to ten times without altering image quality. Dr. Christoph Tillmans from diagnostikum Berlin clinic who is engaged with Siemens Healthineers says that they are regularly using cardiac cine, cardiac MRI at diagnostic resolution even for patients having cardiac arrhythmia with free breathing with the help of compressed sensing. According to Dr. Francois Pontana from University Hospital of Lille, France, CS-MRI significantly improves the visualization of cardiac MRI [10].

GE Healthcare has developed CS enabled MR imaging technology called “HyperSense”. According to them CS can benefit in three ways- reduce scan time, increase spatial resolution, and increase volume coverage. They have demonstrated that CS-MRI can provide 3D knee images in half the conventional acquisition time [22].

6.2.2 Magnetic Resonance Spectroscopy Imaging

Magnetic resonance spectroscopy Imaging (MRSI) noninvasively provides valuable information about biochemical processes inside the human body. It has the ability to detect the metabolic changes within the target voxel. It is commonly used to characterize tissues for the diagnosis of cancer, aggressiveness of tumor, stroke, and Alzheimer's disease. It takes generally longer scan time compared to conventional MRI as it consists of a series of tests. Basically, it measures different types of metabolites, like, amino acids, lipids, lactate, choline, alanine, N-acetyl aspartate, etc. in the target tissues to distinguish disorders. These metabolites are measured in terms of parts per million (ppm) and their response peaks are plotted [15, 30, 41].

In [14], authors demonstrate that it is possible to reduce acquisition time up to 80% in MRSI without significant loss of clinical information. Park et al. [33] achieved four times enhancement in spatial resolution in spectroscopic imaging using the CS technology. It enabled the radiologist a way to visualize heterogeneity in the metabolic profiles within brain tumor tissue.

6.2.3 Diffusion-Weighted MRI

Diffusion-weighted MRI or diffusion-weighted imaging (DWI) has a number of significant applications in clinical body imaging. Most commonly used for tumor detection and characterization, monitoring of treatment progress, etc. DWI signal is generated from the motion of water molecules in the intracellular, extracellular, and intravascular spaces inside the body [23]. Whole-body DWI is relatively new technology and capable of giving unique and valuable information regarding therapy response, bone marrow assessments, etc. [32]. Clinical applications of whole-body DWI for cancer patients are exponentially growing on. It is also used for early detection of breast cancer.

Motion artifacts is a challenge in clinical practice of DWI, like, breath-hold abdomen imaging in whole-body MRI. Application of CS in DWI can significantly reduce the motion artifacts and also improve the image quality. Bilgic et al. [2] successfully reduce the data acquisition time of DWI from 50 to 17 min without compromising image quality. Reconstructed images are compared with original fully sampled data, scanned in 50 min with ten times averaging (considered as gold standard), and shown that RMSE value is sufficiently less. Huang et al. [19] propose an efficient CS-based diffusion weighted MR image reconstruction technique. The reconstruction process explores intra and inter image correlation with the help of joint-sparsity and low-rank regularizations.

6.2.4 *Magnetic Resonance Angiography for Body Imaging*

Magnetic resonance angiography (MRA) produces MR images with highlighted blood vessels; mainly for evaluation of arteries of heart, lungs, kidneys and legs. In clinical practice, it is performed to identify abnormalities of arteries, like, abnormal narrowing, dilatation of vessel wall, etc. Recent developments in MRI hardware like availability of 3.0 Tesla scanner has further made it possible to acquire images with good spatial resolution with shorter scan time, low dose of contrast agent [18]. In most cases, MRA includes contrast agents, particularly gadolinium to shorten the T1 time of blood. Physician advises to perform MRA if patient is suffering from any of the following problems—stroke, heart disease, vasculitis, aortic aneurysm, atherosclerosis, and mesenteric artery ischemia. They are mainly three types— flow-dependent angiography, phase-contrast angiography, and flow-independent angiography. Data can be acquired either in 2D or 3D.

In case of direct data acquisition a huge amount of data need to acquired. In this particular case, integration of CS-MRI helps to improve image quality, data acquisition time, and patient comfort. Rapacchi et al. [36] achieved an acceleration up to 12.5x for contrast-enhanced MR angiography with diagnostically acceptable spatial and temporal resolutions. They carry out this study on six healthy volunteers and obtained similar quality of images to those obtained from fully sampled data.

Conventional non-contrast whole-heart coronary MRA takes significantly longer scan time. Recently, Nakamura et al. [29] demonstrate the advantage of CS for clinical whole-heart coronary MRA. They have successfully reduced the scan time to 3 min 45 s from 15 min 6 s of the conventional approach by maintaining the same diagnostic image quality.

6.3 MRI in Neuroinformatics

Traumatic brain injury is a very common problem. MRI is generally performed after 72 h of injury because initially CT is preferred after brain injury. MRI is better for detection of hematomas, axonal injury and subtle neuronal damage than other imaging modalities. However, for bony pathology CT gives superior performance [24]. MRI is more sensitive to neuronal damage. The development of new technology like, compressed sensing, to reduce MRI scan time may suppress the application of CT for brain imaging in near future.

6.3.1 *Brain MRI*

In clinical applications, brain is the most commonly scanned part of the human body. Brain MRI is performed mainly for detecting the nature of abnormalities and its accurate location. The localization of an abnormality is very important because the same abnormality at different locations of the brain may lead to completely different outcomes, like, the same stroke lesion at different locations could lead to language, sensory, or motor disability. A healthy human brain have more than 10^{11} neurons and each of them is responsible for an unique task. Therefore detection, quantification and localization of structural damage is very essential for analysis of brain via imaging. Exact localization of abnormalities is the most difficult challenge in clinical practice [27]. Generally, T1-weighted MRI is used for accurate localization which is able to provide the highest resolution images, typically, 1 mm in resolution. High resolution imaging requires more number of raw k-space samples which involves long data acquisition time.

Recently Deka et al. [7] proposed an efficient CS-MRI reconstruction method for reconstruction of brain and other MR images from highly undersampled data. They use wavelet domain hidden Markov tree (HMT) model to detect wavelet domain support of MR images from highly undersampled Fourier (k-space) data for the reconstruction of brain and other MR images.

Dynamic contrast enhanced brain MRI is widely performed in clinical practice for analysis of blood–brain barrier (BBB) leakage in brain tumors, epilepsy, migraine, and neuropsychiatric disorders. It requires paramagnetic contrast agent and rapid data acquisition to track the the contrast through the target volume. Due to the slow acquisition, conventional MRI images suffer from low resolution. Guo et al. [16] successfully improved the spatial resolution using CS. It helps in correct characterization of abnormal tissues and provides better image quality compared to conventional approach.

6.3.2 *Functional MRI*

Functional MRI (fMRI) is one of the most popular imaging techniques for detection and measurement of neuronal activity in the human brain and spinal cord without injecting any contrast agent. It gives unique and important information of brain activity and how normal activity disrupted due to disease. Oxygenated blood increases in capillaries of associated locations in the brain where neuronal activities take place. Due to this MR signal intensity changes which is measured by BOLD-based fMRI. Similarly, perfusion imaging is also used to evaluate brain functioning using functional and metabolic parameters. For example, cerebral perfusion which gives blood flow information in brain's vascular network. It has a number of applications in diagnosis of patients having brain disorders. Exogenous tracers like iced saline solution, radionuclides, paramagnetic contrast materials, magnetically labeled blood, etc., are generally used during imaging. It is mainly performed for post-analysis of an acute stroke, detection of Alzheimer's disease, assessment of brain tumors, and effects of drugs [35].

Unlike the electroencephalography (EEG) which is used to detect brain activity from the skull's surface, fMRI measures brain activity from the inside of the brain. In comparison with the positron emission tomography (PET) where radioactive elements are injected in the body and then trace their flow, the fMRI is safer and comfortable. Applications and research of fMRI are continuously growing because there is no other technology which can suppress the ability of fMRI for acquiring information regarding brain activity [38].

Generally, fMRI signals contain noise. Tesfamicael and Barzideh [40], proposed a Bayesian frame based CS-MRI reconstruction method with sparsity and clusterdness priors. It effectively reconstructs fMRI data with better image quality. Fang et al. [9] proposed a CS-based functional MRI reconstruction method and achieved resolution improvement by six folds. Similarly Han et al. [17] experimentally demonstrate that for high-resolution fMRI CS with non-EPI sequence is a good solution.

6.3.3 Diffusion Weighted Brain MRI

Diffusion weighted MRI is used to measure motion of water molecules in brain tissues, like, CSF and white matter. It is an essential neuroimaging technique routinely used for brain attacks and acute lesions. It is very sensitive and gives specific diagnosis details about ischemic and infarction due to stroke [28]. Additionally, diffusion-tensor imaging (DTI) shows the direction of diffusion in a particular voxel.

Fast spin echo (FSE) and Echo-planar imaging (EPI) in parallel MRI significantly reduce the scan time. Reduced scan time provides better image quality by reducing motion artifacts and blurring when EPI is used for DW-MRI. The resolution and signal loss in EPI can be improved by the implementation of CS-MRI. High Angular Resolution Diffusion Imaging (HARDI) enable characterization of complex white matter micro-structure with high precision. But, HARDI require prohibitively more number of samples than traditional DWI, which results in significantly longer scan time. Cheng et al. [5] with the help of CS achieve 6D DWI reconstruction using 11 times less samples compared to fully sampled data with very low RMSE. Duarte-Carvajalino et al. [8] proposed a multi-task Bayesian CS-based diffusion-weighted MR image reconstruction technique. They have used statistical modeling of diffusion signals to explore redundancy and also experimentally validated their model using in vivo DWI data.

6.3.4 Magnetic Resonance Angiography for Brain

Conventional clinical MRA of brain is performed for the assessment of blood supply in various regions of brain. MRA also gives valuable information about shape, size, orientation, and location of vessels in brain. Neuroradiologist used brain MRA to detect abnormalities like widening and ballooning of vessels, brain injury and congenital defects. Generally, 3D time-of-flight (TOF) MRA sequence is used to evaluate arterial blood supply of brain. TOF MRA sequence provides high SNR

MRA without contrast agents. In many cases maximum intensity projection is used as a postprocessing technique to obtain a 2D projection image from 3D data.

Yamamoto et al. [43] demonstrate that using TOF-MRA sequence it is possible to reduce the data acquisition time three to five times using CS-MRI without loss of information. They are able to study image data from patients having moyamoya disease with this incredibly short scan time. Recently, in [1] authors proposed a novel technique for optimization of regularization parameters in CS-based angiography image reconstruction. The results are evaluated using different statistical imaging metrics to reflect radiologists' visual evaluation.

6.4 Commercial CS-MRI Scanners

1. Siemens Healthineers licensed the CS technology for MRI. In February 2017, the Food and Drug Administration (FDA), USA has approved CS-MRI technology for clinical practice. CS-MRI equipped scanner can perform cardiac imaging in just 25 s with free breathing whereas conventional MRI takes more than 4 min with 7–12 times breath holds. Due to the significant reduction in the amount of data acquisition, now abdominal MRI can be performed with free breathing on more number of patients [31]. Recently, Siemens Healthineers has launched CS enabled 3T scanner MAGNETOM Vida [26]. It provides free-breathing and high-resolution abdominal and cardiac MRI with less scan time.
2. GE Healthcare in Autumn 2016 introduces CS technology equipped MRI system called HyperSense, which can perform at significantly shorter scan times without loss of diagnostic image quality. Same technology was also made available in their new 1.5T, 3.0T and powerful ultra-premium 3.0T MRI scanners [22].
3. Recently, Philips in January, 2018 introduced "Compressed SENSE" which is built by incorporating CS technology with sensitivity encoding (SENSE) algorithm. It can accelerate all clinical 2D and 3D MR applications up to 50% without loss of diagnostic image quality [13].

6.5 Conclusions

Recent developments in CS-MRI technology have been able to overcome the main drawback of conventional MRI. At present, well-known MRI imaging system manufacturers including Siemens Healthineers, GE Healthcare, and Philips have already incorporated the CS technology to their state-of-the-art systems. It is expected that in the near future all clinical MRI systems would able to scan 3D and higher dimensional MRI applications within significantly less scan time.

Adoption of CS technology results in the paradigm shift of traditional MRI. Currently, bioinformatics and neuroinformatics have seen tremendous growth in the application of CS-MRI technology for clinical practice and results are very encouraging and best in the industry. Very soon CS-MRI will revolutionize these two fields with its high speed imaging technology and provide new heights to the modern healthcare scenario using noninvasive imaging.

References

1. Akasaka, T., Fujimoto, K., Yamamoto, T., Okada, T., Fushumi, Y., Yamamoto, A., Tanaka, T., Togashi, K.: Optimization of regularization parameters in compressed sensing of magnetic resonance angiography: can statistical image metrics mimic radiologists perception? PLOS ONE **13**(5), 1–14 (2018)
2. Bilgic, B., Setsompop, K., Cohen-Adad, J., Wedeen, V., Wald, L.L., Adalsteinsson, E.: Accelerated diffusion spectrum imaging with compressed sensing using adaptive dictionaries. In: Ayache, N., Delingette, H., Golland, P., Mori, K. (eds.) Medical Image Computing and Computer-Assisted Intervention - MICCAI 2012, pp. 1–9. Springer, Heidelberg (2012)
3. Blasiak, B., van Veggel, F.C.J.M., Tomanek, B.: Applications of nanoparticles for MRI cancer diagnosis and therapy. J. Nanomater. **2013**, 1–13 (2013)
4. Chavhan, G.B., Babyn, P.S.: Whole-body MR imaging in children: principles, technique, current applications, and future directions. RadioGraphics **31**(6), 1757–1772 (2011)
5. Cheng, J., Shen, D., Basser, P.J., Yap, P.: Joint 6D k-q space compressed sensing for accelerated high angular resolution diffusion MRI. IPMI, Lect Notes Comput Sci **9123**, 782–793 (2015). Springer
6. Crasto, C.J. (ed.): Neuroinformatics. Humana Press, New Jersey (2007)
7. Deka, B., Datta, S., Handique, S.: Wavelet tree support detection for compressed sensing MRI reconstruction. IEEE Signal Process. Lett. **25**(5), 730–734 (2018)
8. Duarte-Carvajalino, J.M, Lenglet, C., Ugurbil, K., Moeller, S., Carin, L., Sapiro, G.: A framework for multi-task bayesian compressive sensing of DW-MRI. In: Proceedings of the CDMRI MICCAI Workshop, pp. 1–13 (2012)
9. Fang, Z., Le, N.V., Choy, M., Lee, J.H.: Fang z, van le n, choy m, lee jh. High spatial resolution compressed sensing (hsparse) functional magnetic resonance imaging. Magn. Reson. Med. **76**, 440–455 (2016)
10. Faster MRI scans with compressed sensing from Siemens Healthineers. Siemens Healthineers. https://www.siemens.com/press/en/pressrelease/?press=/en/pressrelease/2016/healthcare/pr. Accessed 29 Jun 2018
11. Friedman, P.D., Swaminathan, S.V., Herman, K., Kalisher, L.: Breast mri: the importance of bilateral imaging. Am. J. Roentgenol. **187**(2), 345–349 (2006)
12. Ganguly D. Chakraborty S., B.M.K.T.: Security-Enriched Urban Computing and Smart Grid. Communications in Computer and Information Science, vol. 78, chap. In: Medical Imaging: A Review, pp. 504–516. Springer, Heidelberg (2010)
13. Geerts-Ossevoort, L., de Weerdt, E., Duijndam, A., van IJperen, G., Peeters, H., Doneva, M., Nijenhuis, M., Huang, A.: Compressed SENSE speed done right. every time. Philips (2018). Accessed 29 Jun 2018
14. Geethanath, S., Baek, H.M., Ganji, S.K., Ding, Y., Maher, E.A., Sims, R.D., Choi, C., Lewis, M.A., Kodibagkar, V.D.: Compressive sensing could accelerate 1H MR metabolic imaging inthe clinic. Radiology **262**(3), 985–994 (2012)
15. Gujar, S.K., Maheshwari, S., Bjrkman-Burtscher, I., Sundgren, P.C.: Magnetic resonance spectroscopy. J. Neuro-Ophthalmol. **25**(3), 217–226 (2005)
16. Guo, Y., Zhu, Y., Lingala, S.G., Lebel, R.M., Shiroishi, M., Law, M., Nayak, K.: Highresolution whole-brain DCE-MRI using constrained reconstruction: prospective clinical evaluation in brain tumor patients. Med. Phys. **43**(5), 2013–2023 (2016)
17. Han, P.K.J., Park, S.H., Kim, S.G., Ye, J.C.: Compressed sensing for fMRI: Feasibility study on the acceleration of non-EPI fMRI at 9.4T. BioMed. Res. Int. 1–24 (2015)

18. Hartung, M.P., Grist, T.M., Francois, C.J.: Magnetic resonance angiography: current status and future directions. J. Cardiovasc. Magn. Reson. **13**(1), 1–11 (2011)
19. Huang, J., Wang, L., Chu, C., Zhang, Y., Liu, W., Zhu, Y.: Cardiac diffusion tensor imaging based on compressed sensing using joint sparsity and low-rank approximation. Technol. Health Care: Off. J. Eur. Soc. Eng. Med. **24**(2), S593–S599 (2016)
20. Kasabov, N.K. (ed.): Springer Handbook of Bio-/Neuro-Informatics. Springer, Heidelberg (2014)
21. Kherlopian, A.R., Song, T., Duan, Q., Neimark, M.A., Po, M.J., Gohagan, J.K., Laine, A.F.: A review of imaging techniques for systems biology. BMC Syst. Biol. **2**(1), 1–18 (2008)
22. King, K.: HyperSense enables shorter scan times without compromising image quality. GE Healthcare (2016). Accessed 29 Jun 2018
23. Koh, D.M., Collins, D.J.: Diffusion-weighted MRI in the body: applications and challenges in oncology. Am. J. Roentgenol. **188**, 1622–1635 (2007)
24. Lee, B., Andrew, N.: Neuroimaging in traumatic brain imaging. NeuroRx **2**(2), 372–383 (2005)
25. Lustig, M., Keutzer, K., V.S., : The Berkeley Par Lab: progress in the parallel computing landscape, chap. In: Introduction to Parallelizing Compressed Sensing Magnetic Resonance Imaging, pp. 105–139. Microsoft Corporation (2013)
26. MAGNETOM Vida embrace human nature at 3T. Siemens Healthcare. https://www.healthcare.siemens.co.in/magnetic-resonance-imaging/3t-mri-scanner/magnetom. Accessed 29 Jun 2018
27. Mori, S., Oishi, K., Faria, A.V., Miller, M.I.: Atlas-based neuroinformatics via MRI: harnessing information from past clinical cases and quantitative image analysis for patient care. Ann. Rev. Biomed. Eng. **15**, 71–92 (2013)
28. Moseley, M.E., Liu, C., Sandra Rodriguez, B., RT(R)(MR), Brosnan, T., : Advances in magnetic resonance neuroimaging. Neurol. Clin. **27**(1), 1–24 (2009)
29. Nakamura, M., Kido, T., Kido, T., Watanabe, K., Schmidt, M., Forman, C., Mochizuki, T.: Non-contrast compressed sensing whole-heart coronary magnetic resonance angiography at 3T: A comparison with conventional imaging. Radiology **104**, 43–48 (2018)
30. Novotny, E., Ashwal, S., Shevell, M.: Proton magnetic resonance spectroscopy: An emerging technology in pediatric neurology research. Pediatr. Res. **44**, 1–10 (1998)
31. New compressed sensing technology could reduce MRI scan times. Rice University (2017)
32. Padhani, A.R., Koh, D.M., Collins, D.J.: Whole-body diffusion-weighted MR imaging in cancer: current status and research directions. Radiology **261**(3), 700–718 (2011)
33. Park, I., Hu, S., Bok, R., Ozawa, T., Ito, M., Mukherjee, J., Phillips, J., James, C., Pieper, R., Ronen, S., Vigneron, D., Nelson, S.: Evaluation of heterogeneous metabolic profile in an orthotopic human glioblastoma xenograft model using compressed sensing hyperpolarized 3D ^{1}3C magnetic resonance spectroscopic imaging. Magn. Reson. Med. **70**(1), 33–39 (2013)
34. Pernet, C.R., Gorgolewski, K.J., Job, D., Rodriguez, D., Whittle, I., Wardlaw, J.: A structural and functional magnetic resonance imaging dataset of brain tumour patients. Sci. Data **3**, 1–6 (2016)
35. Petrella, J.R., Provenzale, J.M.: MR perfusion imaging of the brain. Am. J. Roentgenol. **175**(1), 207–219 (2000)
36. Rapacchi, S., Han, F., Natsuaki, Y., Kroeker, R.M., Plotnik, A.N., Lehrman, E., Sayre, J., Laub, G., Finn, J.P., Hu, P.: High spatial and temporal resolution dynamic contrast-enhanced magnetic resonance angiography (CE-MRA) using compressed sensing with magnitude image subtraction. J. Cardiovasc. Magn. Reson. **15**(1), 1–3 (2013)
37. Rubin, D.L., Greenspan, H., Brinkley, J.F.: Biomedical Informatics, fourth edition edn., chap. In: Biomedical Imaging Informatics. Computer Applications in Health Care and Biomedicine, pp. 285–327. Springer, London, Heidelberg, New York (2014)
38. Smith, K.: Brain imaging: fMRI 2.0. Nature **484**, 24–26 (2012)
39. Symms, M., Jager, H.R., Schmierer, K., Yousry, T.A.: A review of structural magnetic resonance neuroimaging. J. Neurol. Neurosurg. Psychiatry **75**(9), 1235–1244 (2004)
40. Tesfamicael, S.A., Barzideh, F.: Clustered compressed sensing in fMRI data analysis using a bayesian framework. International Journal of Information and Electronics Engineering **4**(2), 1–7 (2014)

41. Tognarelli, M., J., Dawood, M., I.F. Shariff, M., P.B. Grover, V., M.E. Crossey, M., JaneCox, I., D. Taylor-Robinson, S., J.W. McPhail, M., : Magnetic resonance spectroscopy: Principles and techniques: Lessons for clinicians. Journal of Clinical and Experimental Hepatology **5**(4), 320–328 (2015)
42. Toledano-Massiah, S., Sayadi, A., de Boer, R.A., Gelderblom, J., Mahdjoub, R., Gerber, S., Zuber, M., Zins, M., Hodel, J.: Accuracy of the compressed sensing accelerated 3D-FLAIR sequence for the detection of MS plaques at 3T. AJNR. American journal of neuroradiology 1–5 (2018)
43. Yamamoto, T., Okada, T., Fushimi, Y., Yamamoto, A., Fujimoto, K., Okuchi, S., Fukutomi, H., Takahashi, J.C., Funaki, T., Miyamoto, S., Stalder, A.F., Natsuaki, Y., Speier, P., Togashi, K.: Magnetic resonance angiography with compressed sensing: An evaluation of moyamoya disease. PLoS ONE **13**(1), 1–11 (2018)

Printed by Printforce, the Netherlands